To kerry
from Rick

more wine!!

The Role and Limitations of Technology in U.S. Counterinsurgency Warfare

The Role and Limitations of Technology in U.S. Counterinsurgency Warfare

RICHARD W. RUBRIGHT

POTOMAC BOOKS

An imprint of the University of Nebraska Press

Library of Congress
Cataloging-in-Publication Data
Rubright, Richard W., 1974–
The role and limitations of technology
in U.S. counterinsurgency warfare /
Richard W. Rubright.
pages   cm.
Includes bibliographical references and index.
ISBN 978-1-61234-675-5 (cloth: alk. paper)
ISBN 978-1-61234-676-2 (pdf)
1. Counterinsurgency—United States.
2. Counterinsurgency—Technological
innovations. 3. Tactics. I. Title.
U241.R82 2014   355.02'180973—dc23
2014032923

Set in Adobe Caslon Pro by Renni Johnson.
Designed by Rachel Gould.

Contents

Preface

This work started to form in my mind during the 1990s when I was first exposed to the fervor in military circles of the coming revolution in military affairs. As an avid technophile I was awestruck by the potential of technology to profoundly change military capability. Unfortunately, at that time I had not yet studied strategy. Yet as a soldier in U.S Army Special Forces, I was acutely aware that war had a human dimension beyond the realm of photogenic widgets of military hardware.

As the years progressed and the wars in Afghanistan and Iraq became the national focus, I started to notice a disconnect in the U.S. military's conception of technology and its applicability in counterinsurgency strategy. Those parts of the military that did not understand strategy, much less counterinsurgency strategy, were still strident in their belief that technology would prove decisive. I was not alone in this appraisal, and soon a backlash about technology occurred. Yet rather than developing a coherent middle ground of blending technology and strategy, rival camps emerged.

This book started as a doctoral dissertation to address the gap in understanding technology and strategy within counterinsurgency warfare not only in the present but the future as well. I endeavor to make sense of technology and counterinsurgency through a strategic lens; however, to do so I also have to accept the changing nature of technological advancement and how its exponential rate of change is nonlinear. I have found the work as challenging as it has been rewarding.

The book is not meant to be a prescriptive step-by-step guide on how technology is applied to counterinsurgency. Instead the reader should find a metanarrative that facilitates thinking about the application of technology to counterinsurgency warfare. Although how the United States conducts counterinsurgency will change, with commensurate changes in technology, this book should not be limited by time. It should stay applicable as long as people remain the center of gravity in counterinsurgency warfare.

In writing this book I would like to thank Colin Gray and Dale Walton, who challenged me at every turn and always had time. I also want to thank Maggie Peterson, whose patience with my grammar has been stoic.

I reserve special thanks for Sharon and Earl Rubright, devoted parents who pushed me to succeed and who duly supported my efforts to do so. This work would have been impossible without them. Without my father I would never have started upon the path I tread, and without my mother I would never have made it to this point. For my parents my love knows no bounds, and their support is clearly the same.

To all the people who contributed their time to talk to me on this topic, I owe you my gratitude: Tony Nelson; Col. Michael "Coyote" Smith; Col. Stephen Latchford; all of the officers and enlisted personnel of the Fourth Brigade, Second Infantry Division, but especially Captains Destructo and Contact; and many, many more. And, of course, thanks for the loving support of my extremely tolerant wife, Nichole.

Last I want to acknowledge the infantry. They are men and women willing to face the long odds when needed. No matter when or for what cause they stood, this book is for them. Usually the outcasts or poor, they are people who know their own worth even if others do not. In these soldiers resides the best and worst of mankind, and it will ever be so.

The Role and Limitations of Technology in U.S. Counterinsurgency Warfare

Strategy and Context

The advent of the computer and the microprocessor set off the Third Industrial Revolution. Its impact has been equally far-reaching. If anything, the pace of change has accelerated.
MARSHALL GOLDMAN

I. Definitions and Terms

It is most helpful to start by defining the terms that will appear throughout this book. While U.S. military intervention may take on an appearance of conventional or unconventional form, or a combination of the two, this work is most concerned with counterinsurgency warfare. This foundational context of future war does not imply that the United States will never again face a conventional or peer competitor on the field of battle; rather, the United States is more likely to face the challenge of counterinsurgency warfare in the continuing battle against opponents who are driven by ideology based on religious beliefs or deeply held issues of identity.

Counterinsurgency warfare has had many names. While this author prefers simply using the term "warfare," in today's lexicon it could be described as low-intensity conflict, guerrilla warfare, revolutionary warfare, and unconventional warfare. Each of these terms has specific connotations, such as "guerrilla warfare" referring to hit-and-run tactics, but attempting to use each term separately to highlight nuances would become unwieldy.[1] In this work the term "counterinsurgency warfare" is used with an implied caveat that the United States from time to time may conduct insurgency warfare, which is referred to as unconventional warfare in modern doctrine. Every term used to describe counterinsurgency warfare has shortcomings. I have opted to use counterinsurgency warfare because it best describes the present challenges that the U.S. military faces on the battlefield. Counterinsurgency also has the advantage of being popular in U.S. policy-making and military circles and thus benefits from wide recognition just as the previous term "low-intensity conflict" did in its time.[2]

The term "counterinsurgency warfare" should not be taken to imply that every conflict in every geographical region in which guerrilla forces are active is exactly the same. The term is only meant to describe an all-inclusive concept of a type of conflict that the United States may feel compelled to enter. Each geographical region, with different languages, customs, people, and religions, may vary widely and require a wholly different strategy to attain U.S. policy objectives. The term "counterinsurgency" thus simply describes the type of warfare the United States is likely to engage in in these disparate regions. The term also implies the United States supports the anarchical state system of global politics and its continued prominence.[3] A counterinsurgency effort, by definition, fights an insurgency for control of a sovereign political entity that the insurgency seeks to supplant. As such, almost assuredly U.S. military intervention will be waged in support of states rather than the opposite. Exceptions to this, most notably, are the U.S. government's support for the insurgencies of the contras in Nicaragua and the mujahideen of Afghanistan in the 1980s.[4]

In this book, the term "counterinsurgency forces" is used to describe both host nation and U.S. forces, possibly with allies, working in conjunction. As a general rule, the United States, due to its own cultural imperatives, almost always wants the host nation forces, which it supports, to establish a sovereign government or to retain its sovereignty within the geographical area in question and in the proper context of a state.[5] Therefore, it follows that the forces supported by the United States, within the definition of U.S. policy objectives, will represent support for an organization with a legitimate claim to becoming a national government and, thereby, a host nation. By definition the forces of a host nation cannot be domestic insurgents; therefore, U.S. military forces working together with the host nation forces will be fighting in a counterinsurgency role.

This book uses the term "revolution in military affairs" to describe the upsurge in technical capability and doctrinal change that has been the subject of intense study and writing since the late 1990s and is addressed in detail later.[6] While some argue whether a revolution in military affairs has taken place, is taking place, or will take place, that debate is not pertinent to this book. Further, using the term is not an attempt to discard the concept of the military technical revolution. Rather, the term "revolution in military affairs" is a broad overarching theme that includes the technological advantages and doctrinal changes available to the U.S. military specifically in terms of firepower, mobility, protection, sustainment, command, control, communications, and intelligence.[7]

This book is not an attempt to deny the concept of the military technical revolution, nor does it wish to enter the debate on how many, when, or with what degree of technical nuance doctrinal changes are required for a revolution in military affairs to exist. The term is used here as a matter of convenience and, owing to its widespread use in current vocabulary, to describe increased technical capabilities as Michael Vickers does.[8] Herein the term "revolution in military affairs" is applied to the doctrinal changes that are possible in the U.S. military force structure today because of the increased technical capabilities that have recently become available and because of others that are likely to emerge.

The term "American culture" describes a set of norms, assumptions, and mechanisms that influence U.S. policy. Clearly any attempt to define American culture within one context or through a single definition is doomed to failure. The United States is so diverse that no one context is applicable. However, providing a helpful narrative are some dominant themes throughout American society and U.S. history, such as the expectation that the government will consider the will of the citizens.[9] As limited as any definition of the term "American culture," with inevitable pitfalls, it remains the best overall concept to help clarify some issues in this book, especially as it pertains to the constraints on U.S. military capabilities.

Finally the operationally offensive, tactically defensive concept is explored as it is directly related to the U.S. military. The concept's dynamics and specific issues related to American soldiers are discussed in chapter 3 onward after the concept itself is defined and historically contextualized. This book recognizes that the capabilities and cultures of soldiers will always play a part in any doctrine.

II. The Macro Question

What is the most appropriate role of technology in U.S. counterinsurgency operations and how does the United States best leverage its technical superiority to ensure harmony between military capability and sound strategic practice? This question is easy to understand but extremely difficult to answer. Not only is technology ever changing, but also the political goals that strategy pursues are likewise often in flux.

Technology provides enormous opportunities to enhance military capability, which stands as one of the most important facets of strategy. Relying on technology alone is not enough for a force to prevail in a counterinsurgency conflict environment, as technology is an enabler for a plan but not in itself a substitute for one.[10] The counterinsurgency environment is so varied and fundamentally different from a force-on-force conventional conflict that technology, while extremely powerful, cannot solve most of the issues facing the United States in confronting ideologically driven opponents. The United States today can field numerous technical capabilities to confront and defeat threats in theaters around the world, but counterinsurgency warfare by its very nature presents further challenges.

Technology can greatly enhance the likelihood of success in a counterinsurgency effort. A common conception, however, is that technology and its advantages can become detriments to the counterinsurgency forces. For example, technology and its advantages lead to increased counterinsurgent logistics, thus providing opportunities for an adversary to acquire supplies.[11] Firepower is also often singled out as doing more harm than good as it tends to alienate the civilian population, which is inevitably vulnerable in warfare.[12] Cheap and inexpensive improvised explosive devices can disable and kill vehicles and soldiers, which are financially and politically expensive.[13] The opponents of technology, meanwhile, fail to focus on its potential when applied to sound counterinsurgency doctrine. Expensive tanks being destroyed by inexpensive devices do not illustrate a failure of technology in counterinsurgency warfare. Rather, using such large, heavy, logistically expensive, predictably moving, and manpower-intensive platforms as tanks in counterinsurgency operations is simply an extremely poor application of technology to the strategy of counterinsurgency operations.

The American concept of warfare itself is also limited in its ability to deal with counterinsurgency warfare. While some aspects of American culture lend themselves to confronting some of the challenges of counterinsurgency conflict, such as establishing institutions that foster justice, elect a representative government, and support the expressed will of the people, other aspects—specifically, the U.S. military's lack of strategic education and its organizational culture, which has been resistant to adaptation and learning—make Americans profoundly unsuited for providing effective support to a counterinsurgency effort.[14]

Counterinsurgency warfare can drag on for years as the battle is waged for the minds and allegiance of the populace.[15] It not only requires tremendous resources but also demands patience and commitment, which are not hallmarks of U.S. support for military endeavors. Over the course of a counterinsurgency conflict, it is often hard to gauge success by any numerical measure, easily explained time lines, or other measurements that are likely to give continuous feedback to indicate and evaluate success or failure in the endeavor.[16] A nation's fundamental inability to measure whether it is "winning" or "losing" in a conflict generates an inevitable debate as to whether continued operations are justified given their often expensive and long-term natures. The United States, having a democratically elected representative government, will always find itself questioned when participating in counterinsurgency conflicts. Inevitably it will result in attempts to minimize casualties and to being risk averse in military operations.

Counterinsurgency efforts typically are longer-term endeavors and are quite different from the conventional wars of past centuries. People who view warfare as inherently evil endeavors that should be banned cannot appreciate the profound and obvious differences between conventional warfare and counterinsurgency warfare.[17] The typical result is the third-party attempt to apply rules and norms from past conflicts to present and different ones and to ignore the realities of the conflict being fought.[18] The constraints on warfare, especially when related to the applications of technology, that stem from third-party interpretations of international laws represent a direct threat to the ability of the United States to pursue national political objectives through counterinsurgency warfare.

III. The Theoretical Context

The existing counterinsurgency literature is devoid of an analysis of the appropriate role of technology and strategy. This book intends to fill that specific gap. Doing so requires a degree of context based on counterinsurgency literature of the past century. This book does not aim to dispute the general conclusions of past writers such as David Galula, Robert Thompson, Robert Taber, David Kilcullen, John Nagl, and David Petraeus; rather, these writers' works form the foundation from which this book draws conclusions about the appropriate role of technology in U.S. counterinsurgency operations.

It is also prudent to make clear that the work is taking a noncanonical approach to counterinsurgency literature. While writers such as Santa Cruz de Marcenado have concerned themselves with counterinsurgency prior to the start of the twentieth century, their insights are not relevant to this book's focus on contemporary counterinsurgency operations conducted by the U.S. military. The noncanonical approach is necessary as the subject matter of counterinsurgency operations is an ever-evolving form of military operations with a tremendous number of variables unique to each conflict. Sun Tzu and Carl von Clausewitz are universally relevant, as is Mao Zedong, because they deal with the nature of war and a tried blueprint for insurgencies; however, no such canon exists for counterinsurgency practices. This book examines the major writers on counterinsurgency of the last hundred years as they have the unifying theme of winning "hearts and minds." Beyond these writers, the value of additional textual analysis offers diminishing returns almost to the point of irrelevance.

This book is not directly concerned with the theories of international relations; however, it is helpful to provide a context in which assumptions are being made. Given the preponderance of power that it has in both the economic and the military realms, it is likely, but not assured, that the United States will continue to act in a realist tradition when dealing with international issues. The United States will pursue its national goals, based on the assumption that they exist, within a system of states in which no amount of optimism or hope will change the nature of endemic violence.[19] However, a purely Hobbesian or Machiavellian moral and legal vacuum does not exist. It should be stressed that U.S. foreign policy emanates not from a legitimate international order but from a domestic U.S. conception of the importance of morality when pursuing policy objectives. While it would be very easy and convenient to explain U.S. foreign policy as a purely classical realist exercise of power, there are usually constraints, self-imposed, upon the exercise of that power.

In the realm of the international system of states, it is almost guaranteed that the United States will continue to emphasize the state as the primary and legitimate political unit of importance. This perspective is particularly important when considering the likelihood of U.S. military intervention. Anytime the United States deploys abroad it must take into consideration the ramifications of the system of states and how its military operations will impact and be impacted upon by the states affected. Low-intensity conflict and guerrilla war, as noted earlier, are more apt to be described as counterinsurgency operations precisely for that reason; the United States will almost always be defending a state (provided the state has democratic tendencies) or supporting the formation of a state when it intervenes militarily. It will almost always be fighting against insurgents who choose to supplant a political organization within the international system of states. Such concepts support the traditionally defensive posture of U.S. strategic thinking with American cultural proclivities.

This book is more concerned with the theoretical constructs and classifications of strategy as an amalgamation of Clausewitz and Sun Tzu than a theoretical context within international relations. Ultimately the United States, given the preponderance of military and economic power at its disposal, is usually able to choose how to engage with the international community on its own terms. While some administrations may favor a more multilateral approach to international relations, others—as the George W. Bush administration (2000–2008) clearly showed—may prefer a more unilateral approach. While an administration may actively make this choice in its outlook and behavior toward the international community, the same cannot be said for instituting radical changes in the application of theory to military strategy. For an administration to make radical changes to an established military strategy, as set by a prior administration, it would require an understanding of the subtle nuances and differences in military strategy. Thus such a shift is not impossible but likely to be politically difficult.

While Mao, Sun Tzu, and Clausewitz never specifically wrote about counterinsurgency operations, the three strategists provide the needed context for conducting an insurgency. Sun Tzu and Clausewitz outline the strategic underpinning of any insurgency, precisely because insurgents are fighting a war; Mao offers a practical guide as to how to actually implement an insurgency. Sun Tzu's emphasis on deception and Clausewitz's emphasis on violence provide a strategic reference not only for insurgents but also for counterinsurgency forces. In short every counterinsurgency force should already know the strategic underpinnings of an insurgency because they are the same for every war fought. Mao's three stages of revolutionary warfare serve as a roadmap for insurgents, but they also allow the counterinsurgent to understand the relative position of the insurgency and act accordingly. Taken together, Sun Tzu, Clausewitz, and Mao offer universal principles to guide both insurgents and counterinsurgency forces.

That Sun Tzu and Clausewitz are still studied indicates the timeless quality of their contributions to understanding the nature and conduct of war. The violence of war, the importance of determining the center of gravity and tailoring strategy to target its specific enemy, the tactical lessons of striking at an enemy's weakness, and the use of deception will forever be a part of war as any practitioner can attest. More contentious could be the claim that Mao's principles of revolutionary warfare are equally applicable and timeless, yet recent history does support such an assertion. Successful insurgencies in China, Vietnam, Algeria, and Nicaragua, as well as ongoing insurgencies in Afghanistan, Peru, Turkey, and Colombia, all support Mao's blueprint.

In the realm of counterinsurgency warfare, both Sun Tzu and Clausewitz have dated but profound theoretical and strategic contributions to make. The theoretical contexts of Sun Tzu and Clausewitz cannot provide a prescription for certain success in a counterinsurgency conflict environment; nor should they be expected to do so. As noted this work argues for an amalgamation of the theories of Sun Tzu and Clausewitz, together with a study of Mao's practices, to provide the insights required for better understanding and prescribing an effective response to counterinsurgency challenges. The United States now faces insurgencies, and likely will face more in the future, but it does so at a time when, backed by sound theory, technology can provide increased military capabilities for better counterinsurgency practice.

This book is not concerned with the writings of Antoine-Henri Jomini. For an effective counterinsurgency strategy, Jomini is likely to be counterproductive as his emphasis is rarely about force-on-force engagements or interior lines.[20] The amalgamation of Sun Tzu and Clausewitz yields the best possible theoretical framework for examining the requirements of counterinsurgency warfare as it presents both the subtle and the unsubtle methods required for dealing with very complex and varied conflicts. In such conflicts, Sun Tzu argued, fighting battles is not necessarily indicative of strategic success and often is not the mark of a great general:

III. ATTACK BY STRATAGEM

1. Sun Tzu said: In the practical art of war, the best thing of all is to take the enemy's country whole and intact; to shatter and destroy it is not so good. So, too, it is better to recapture an army entire than to destroy it, to capture a regiment, a detachment or a company entire than to destroy them.

2. Hence to fight and conquer in all your battles is not supreme excellence; supreme excellence consists in breaking the enemy's resistance without fighting.[21]

Yet counterinsurgency also requires that violence be used in order to achieve the political objectives that are set out in U.S. national policy. Clausewitz emphasizes this stipulation: "Violence arms itself with the inventions of Art and Science in order to contend against violence. Self-imposed restrictions, almost imperceptible and hardly worth mentioning, termed usages of International Law, accompany it without essentially impairing its power."[22]

In counterinsurgency warfare, the United States will find that its opponents have political motivation, possibly cloaked in the shadow of ideological or religious fundamentalism, as their primary driver of insurgent will. However, the insurgents' use of force, as Clausewitz would argue, is simply an extension of the political process by which they will attempt to obtain their policy objectives.[23] In such an environment, and as he proved with his successes, Mao provides a theoretical framework to help clarify this point by examining and applying his three stages of revolutionary warfare.[24] While there is no guarantee that adversaries seeking to establish a state as their political objective will follow the theory of Mao's three stages of revolutionary warfare, history shows that such attempts may produce positive results. It is equally clear that all three stages of Mao's revolutionary warfare do not have to be followed in order and therefore are not held as some canon of how an insurgency must unfold. The three stages merely represent a theoretical context that is helpful in highlighting the role of technology and the constraints limiting counterinsurgency conflict.

Counterinsurgency as a practice is warfare by every definition. The distinctions between conventional warfare and unconventional warfare are differences that are wholly linguistic and both helpful and unhelpful. Counterinsurgency by itself recognizes a different battlefield with a different center of gravity from that of conventional warfare, yet their principles remain the same.

At present counterinsurgency practices are not given the free range they had in the past.[25] While counterinsurgency operations are extremely easy to carry out successfully with conventional forces, the scale of slaughter needed to quell an unruly populace is horrific and morally dubious. History is replete with examples of extreme and harsh methods used successfully to subdue insurgencies such as the U.S. campaign against Philippine insurgents.[26] The U.S. military's current practice of counterinsurgency, however, does not focus primarily on mass slaughter.

Clausewitz's identification of enemy centers of gravity, found in the thoughts and allegiance of the population, still applies in counterinsurgency.[27] Therefore, acknowledging and acting upon his theory will cause the enemy to lose the allegiance of the people. As an insurgent force spreads its ideological and political vision, counterinsurgents should realize that it is the people, not battles or body counts, who provide a strategic basis for success. It is difficult to overemphasize how important fostering the loyalty of the people is to success for either side in a given counterinsurgency struggle. Mao Zedong most aptly describes this principle in the overly quoted, yet perfectly stated, maxim that relates the "guerrillas as fish" and the people as the water in which the fish swim.[28] Without the support of the people, the insurgent is doomed to failure, and any strategy for success by the counterinsurgency forces must keep this fact at the forefront of their decision-making process. Yet at the same time, the counterinsurgency forces must be prepared to engage, with extreme violence, those insurgent elements within its grasp.

Clausewitz's view of warfare as a violent duel cannot be disregarded.[29] Counterinsurgency warfare is not just civic action and social engineering.

Counterinsurgency also lends itself to the thinking of Sun Tzu as much as to that of Clausewitz. Not every engagement must be a violent, bloody confrontation or duel, assuming that the current aversion to mass slaughter as a tactic continues. Civilians generally do not like firefights in their own neighborhoods, and in most cases a population will simply hunker down and support whichever side it feels will either win or cause them the least harm.[30] It should be noted that this behavior of most people holds in the majority of cases, but it is far from true in every element of warfare. Understanding the people becomes a paramount factor in assessing whether they have the will to give active support to the insurgents and their cause.[31] In such a situation, where winning the support of the populace without destroying their wealth or well-being constitutes success, a military leader who wins battles without fighting is indeed superior to all others. In short, political accommodation of the people in order to maintain their loyalty to counterinsurgency forces, through deception, cajoling, intimidation, or other nonviolent methods, is a perfectly valid choice provided, of course, that the application of such tactics does not harm the strategic effort.

At present the United States cannot, or will not, subjugate people through mass slaughter or bring homogeneity to a geographical region through genocide of minority groups. Instead, the United States is prepared to compete for the hearts and minds of the people in the conflict zone. The United States has failed to adequately do so until very recently, possibly because of the cultural and institutional proclivities that make the complex world of counterinsurgency strategy difficult. This failure likely stems from a lack of strategic thought in the United States—both in the military and the government—that then makes the preparation of a sound counterinsurgency strategy almost impossible. Additionally to understand how technology will impact counterinsurgency practice, first the complexity of counterinsurgency must be understood or else the application of technical capabilities will be random and ineffective.

There are many successful methods to supplant the ruling body of a state, such as a coup—which relies upon speed, secrecy, and timing—or the more systematic method found in the form of Mao's revolutionary war.[32] With its three distinct phases of defense, stalemate, and offense, Mao's theory represents a historically effective method for successful insurgency operations.[33] Most important as the wholesale slaughter of people is not currently a viable military strategy for the United States, the theory recognizes the significance of securing the support, or at least the neutralization, of the populace. Understanding Mao's three stages of revolutionary warfare is useful for understanding not only counterinsurgency doctrine but also the role of technology in supporting that doctrine. As each phase is distinct from the others, the counterinsurgent must utilize his technology to best serve the appropriate doctrine to meet the challenges of each phase. While Mao would consider the third phase to be part of revolutionary war, owing to its conventional nature in opposing counterinsurgency forces, the third stage would be classified as conventional warfare. This blurring of the conventional and unconventional, recently described as hybrid warfare, should not be an issue for the U.S. military as any attempt by an insurgent group to move to conventional operations plays directly into the strengths of U.S. military power.[34] Further because counterinsurgency warfare is still warfare in its true nature, the ability and expectation of counterinsurgency forces to adapt readily to either conventional or unconventional warfare should be constant.

Of the counterinsurgency writers surveyed in this section, a single common theme runs throughout their works: the hearts and minds of the people must be won, and that process starts with protecting them. Thompson, Galula, Taber, Nagl, Frank Kitson, Kilcullen, James Amos, and Petraeus all recognize that securing the populace and winning their hearts and minds are the only viable ways to effectively counter the influence of an insurgent movement in today's political climate. (Col. C. E. Callwell is the lone exception given the date of his writings.) As discussed further later, each of these writers offers different lessons on counterinsurgency, and they are the foundation upon which the U.S. military will rely as it attempts to enact a counterinsurgency strategy. However, none of these authors presents any useful insights into how technology is used to further sound counterinsurgency strategy or practice.

The current direction of U.S. counterinsurgency doctrine is based on the thinking of David Petraeus, who spells out the fundamental problems of conducting counterinsurgency operations and poses solutions in the 2006 Army field manual *FM 3-24*.[35] The document is historical in nature, borrowing heavily from David Galula, and while it informs some consistent aspects of counterinsurgency, it is dated. Galula's recognition of the populace as the deciding factor in a counterinsurgency's effort and methods, such as taking census and control, are wise counsel.[36] However, Galula (1919–67) was a man of his time, and while his principles are generally correct, his devised methods must be seen in the context of the technical capabilities then available. It follows that the more advanced capabilities of the United States can now alter the method of counterinsurgency operations but that they should be employed while mindful of the political objectives outlined by policy makers.

Other prominent writers in the field of counterinsurgency such as Colonel Callwell had similar contributions to make about the basic nature of counterinsurgency warfare. They are often not applicable in the detailed and micro evaluation of today's methods due to the technical, and at times ideological, gaps that exist between the time when they were written and the current available capabilities. Callwell had sound tactical advice and demonstrates an understanding of the linkage between the tactical effort and a larger strategic military effort when he states, "Columns owed their strength to rapidity of movement rather than to numbers, they kept were kept equipped and supplied with a sufficiency of pack transport, artillery and baggage being reduced to a minimum."[37] Here he provides sound tactical advice for small, mobile insurgent-hunting teams. The idea remains powerful in that a mobile, lethal, and relatively low logistical-cost soldier is an ideal counterinsurgency asset for which new technologies can be developed. However, his view was that "it cannot be insisted upon too strongly that in a small war the only possible attitude to assume is, speaking strategically, the offensive. The regular army must force its way into the enemy's country and seek him out."[38] This approach could not be more wrong, but in the context of counterinsurgency in his day, punishment and reprisals were the norm.

Likewise, ideology is also subject to becoming dated. Frank Kitson's description of insurgents in Oman is an informative account of ideological motivations mixed with attempted personal gain.[39] His account of the land dispute that led to the Mau Mau insurgency in Kenya, with its local tribal customs and idiosyncrasies, is interesting but hardly on the same level of virulence as al-Qaeda's mass-murdering international jihad against all things Western.[40] Yet Kitson does offer valuable and timeless lessons, such as the recognition that insurgents will always have a personal angle to their political motivations and, more important, that good intelligence collection is crucial in driving effective non-kinetic and kinetic counterinsurgency operations.[41]

Both Robert Thompson and Robert Taber wrote in a time of burgeoning communist insurgencies spreading throughout the world. Based on Thompson's Malayan experience, his recognition that the people are of primary value, and their isolation from the insurgents essential, is timeless.[42] So too is Taber's recognition of the fundamental aspect of the will of the insurgents.[43] However, dating again becomes an issue as both men were focused on the communist threat of their time and were required to work within the technical framework that they understood. While the communist ideology of the 1960s may have had a powerful attraction to many, Vietnam should have shown that there are far deeper identities than imported political and economic ones. The methods and views of insurgents and, more important, their appeal to local populations based on a foreign economic ideology are nowhere nearly as powerful as those based on a tribal or religious identity, such as those found in the Arab world, where the United States may well find itself conducting counterinsurgency operations. Thompson is right when he says that nationalism was the real vehicle for the insurgency in Vietnam, and Taber is right when he focuses upon the will of the people to fight for that nationalism. Counterinsurgency is not as hopeless as Taber makes it seem, however, and Thompson's methods of isolating the populace are not necessarily suited to today's technical and political environment.[44]

John Nagl, among other later writers, does not particularly help the United States move forward with concepts of strategy or technology in counterinsurgency operations as he identifies the U.S. military's historical problem of failing to adapt but does not prescribe an adequate solution. Likewise his view that the U.S. military required more flexibility and strategic adaptability is absolutely correct in relation to Iraq in 2006, as is his appreciation for the adaptability of the British military's counter-insurgency effort in Malaya.[45] However, arguing for adaptability when carrying out counterinsurgency strategy should have been a foregone conclusion in any counterinsurgency effort. The cliché that no plan survives first contact intact is hardly new, nor does it help in understanding counterinsurgency better, but Nagl is right in emphasizing that the U.S. military needs to be a learning and adapting organization. It is perhaps most damning that such a fundamental shortcoming had to be pointed out publicly in order to garner attention.

The examination of the causal effects, organizational structure, and grievances of current insurgent forces is important work and extremely helpful in understanding the opportunities that technology provides for good counterinsurgency practice. David Kilcullen outlines, in an outstanding way, the method al-Qaeda uses to infiltrate villages and exercise control over indigenous populations in order to draw them into its ideologically driven political movement.[46] His observations directly affirm the basic principle that the allegiance of the people matters the most in any counterinsurgency environment, and he provides an analysis of where technology can fit to undermine the insurgents' efforts.

Unfortunately the lack of emphasis on technology in counterinsurgency strategy as an effective tool to enhance good counterinsurgency practice is missing from the literature with a few exceptions. David Tucker, for example, emphasizes that irregular warfare should not be ignored with regard to the technical innovations of the revolution in military affairs.[47] Many technical programs have been tried, such as deforestation in Vietnam and harassing fire from artillery, but the technology has not been incorporated into proper counterinsurgency strategy.[48] Arguably this is partly because the military lacks the appropriate technical capabilities to enhance the abilities of the counterinsurgency forces. Fundamentally it is hard to develop technology for the U.S. military's use that is directed toward winning hearts and minds. The issue has historically seen a misapplication of technology in support of the U.S. military's poor strategy.

Counterinsurgency literature is unfortunately devoid of a serious consideration of technology and the capabilities that may be offered. This absence of technical consideration is a natural outgrowth of the counterinsurgency literature discussed earlier. Winning hearts and minds is simply not a subject matter that easily lends itself to technical considerations. However, given that the first step in winning hearts and minds is in providing security to the populace, a lack of technical consideration becomes more puzzling. Security is heavily influenced by technical capabilities.

The gap in the literature, while unfortunate, is to be expected. Most practitioners of counterinsurgency operations are not versed in rapidly occurring and fading technical trends. Most technically oriented people are not informed in the actual practice of counterinsurgency operations, and strategy is wholly deficient in the United States in general. To begin to answer the research question requires a multidiscipline approach. This method makes addressing the research question unwieldy and feels fragmented at times, but there is simply no other way to address the question if the literature gap is to be filled.

There are a few overarching themes in the literature on technology. Main among them is the notion of a revolution in military affairs that has been brought about by the advent of micro-processing and sensor technology. Seminal works, such as *Breaking the Phalanx* by Douglas Macgregor and *Yellow Smoke* by Robert Scales, are excellent examinations of the possible results of technologic capabilities on conventional force structures. In fact the body of literature is legion, resulting in U.S. military initiatives such as transformation, the Army after Next, net-centric warfare, and the objective force. However, as uninterested as authors on counterinsurgency are about technology, the same is evident about the technophiles' apparent disinterest in counterinsurgency strategy.

A strong argument exists that a revolution in military affairs is currently occurring in the United States as it develops and integrates new technology into its force structure and adapts doctrinal changes to make the best use of the new technical capabilities. There should be little debate about whether this trend is historical or new. The U.S. military has generally been a technically driven organization at least since the Civil War. The current debate is whether the revolution in military affairs constitutes something profoundly different within the sphere of military capability.[49] William Hurley rightly identifies that technology has tremendous potential to allow different approaches to old problems, such as urban operations, but such a debate highlights the tendency in American culture to emphasize the technical capabilities of the U.S. military. Fundamentally there is a perception that military capability will be directly enhanced through the integration of technology. Given the past emphasis on force-on-force projected conflict, such a perception is understandable. However, the perception of enhanced military capability through technology is both a blessing and, at times, a detriment when applied to the realm of counterinsurgency warfare.

Technology is not strategy; rather, it exists in a vacuum until applied to enhance capability. Even as a capability, though, it is still far short of any strategic meaning. Strategy must be made responsive to political objectives at all times. As a strategy is used to seek ways of obtaining political objectives, the development of the strategy must always be mindful of what is reasonably possible to do in pursuit of the objective. As strategy takes into account the seventeen different elements of strategy outlined by Colin Gray, it should be noted that only one of them—a very important one—has to do with military technology.[50] Gray, probably the most balanced writer, correctly identifies the potential of technology while maintaining the need for such techno infatuation to remain properly contextualized in strategic considerations. The military's capabilities directly dictate the military options available to the policy makers. Often the greater the military capabilities of a state, the more numerous and varied are the options available to the policy makers and, in turn, the greater the strategic advantage a nation may possess.

The possible misuse of technology is not an inherent limitation of the technology itself. Technology is benign with a latent potential rather than a substantive meaning or purpose. Technology becomes a detriment through the failure of human strategic conception. Such failure has been abundantly clear in modern historical examples such as the U.S. intervention in Vietnam. There was nothing inherently or morally wrong with any of the technology used in Vietnam at the time. Instead, the failure was in the human factor of how the technology was used to enhance military capability and to further U.S. strategy in the ultimate grand plan of containing communist expansion. In short, technology only becomes a problem when it endangers the harmony between different levels of conflict by steering strategy toward what is easy and comfortable for the existing force structure.

Doctrine applied by a military force is as important as the technical capabilities of the military devices themselves are, as was evident during the initial success of the German blitzkrieg during World War II.[51] The common perception is that the doctrine will drive the military capability. As long as a military sees the likely adversary as a peer competitor, the military capabilities developed will tend to improve upon the current systems and, in turn, further entrench the current doctrine.[52] As long as the doctrine remains in place, development expenditure will mirror the current structural needs that it establishes.

George Friedman and Meredith Friedman do an excellent job at contextualizing technology and its likely trajectory with the attendant opportunities for enhanced force structure. They correctly outline the likely technologies, such as sensors and unmanned aerial vehicles, that will make a radical shift in military capabilities possible. Unfortunately like many technology-oriented writers, they are wholly absent on the subject of strategy.

Acquisition and integration of technology into the force structure can come at the cost of strategic thought and a tendency to ignore complex issues like counterinsurgency warfare. Attractive new technical systems run the risk of emphasizing what the system can do rather than emphasizing what their shortcomings are when used outside a very narrow niche. This observation obviously should not be taken to mean that technical capability within a force structure has an overall negative net effect. It is true that enhanced military capability does provide flexibility and certainly can lower financial and political costs for a force structure. It should not be forgotten, however, that counterinsurgency conflict must be a holistic exercise that brings together both military capability and doctrine to further strategy. The appropriate application of technology to counterinsurgency warfare is a difficult concept to grasp and is nearly impossible when a sound counterinsurgency doctrine is absent.

Almost all of the technical and journal articles that expressly deal with technology are concerned with conventional force structure. Whether it is the technical aspects of Boeing products, such as the joint direct attack munition, or the Lockheed Martin F-22 Raptor or Brett Steele's article on the reengineering during the interwar period, minutiae of specific technical capabilities or very narrow technical questions tend to be considered in isolation. The technical literature deals with questions that are entirely devoid of strategic consideration in favor of force structure analysis or specific military capabilities.

Many challenges to understanding the impact of technology within the force structure of the United States, and to fully understand the potential for enhanced military capability, lie in understanding the nonlinear nature of technical improvement. It is easy to see long linear evolutions of technical capabilities as devices continue to improve within the technical parameters that define them. Any device can be understood to become more capable within its own role as technology is added to provide increased capabilities.[53] The range of main tank guns has increased over time to provide heightened lethality, but such linear progression in the evolution of a military capability is not the whole story. Rather, these examples cloud the issue of nonlinear, or exponential, increases in capability as multiple technologies combine to provide significant new capabilities, such as the Germans' use of radios in tanks at the beginning of World War II.[54] This book puts this issue into perspective and, more important, provides an adequate window to explain the profound importance of technology in counterinsurgency warfare when applied correctly to support appropriate doctrine.

The nature of nonlinear acquisition of technical capabilities further highlights a very important point: there are clearly no crystal balls giving foresight to the role of technology.[55] Extrapolating the future based on history is fraught with peril and will as often as not lead to inaccurate conclusions. Of course the more detailed the prediction the more likely it is to be wrong. Predicting the specific future technical capabilities of the U.S. military is simply not possible. However, it is possible to understand overall trends and to provide a cogent picture of technology as it relates to the U.S. military when engaged in counterinsurgency warfare within an appropriate theoretical context. Clearly some attempt does have to be made to understand the likely trajectory of where technology will lead the

U.S. military in counterinsurgency warfare. But this effort is profoundly different from predicting specific capabilities out to specific time horizons.

While history does provide an outstanding narrative to understand the present situation of U.S. military capability, and its strategic failures and accomplishments, history may not be the omniscient tool for understanding technology or its implications. Clearly while history is our best tool for recognizing trends and likely avenues of technical advancement, it will never provide, nor should it be expected to offer, an absolute frame of reference for how new technologies will be employed. As technical capabilities emerge, some may be wholly new in the history of warfare. This is not to say that they will or even can change the nature of warfare, but new technologies or new technical capabilities do have the potential for changing specific conflicts as, for example, the Stinger missile did in Afghanistan.[56] This book argues that counterinsurgency warfare provides a good environment for new technologies or technical capabilities to be developed and implemented.

This work relies heavily on historical examples to highlight trends in the U.S. military and American culture. The book explains specifically how technology, when applied properly to counterinsurgency conflict—such as through an operationally offensive, tactically defensive doctrine—could provide important new capabilities for the U.S. military and echoes a conventional theme in Robert Scales's work.[57] An operationally offensive doctrine combined with a tactically defensive capability relying on technology is not a new concept. Technology will provide a heightened ability to make this doctrine a reality. This book examines some historical examples of how the technical capabilities will provide future opportunities.

Technology is a diverse term that conveys numerous aspects of military capability. Technology is, in essence, an enabler of capability. It is extremely important that technology not be limited to photogenic platforms suited for kinetic effects. Technology does have considerable abilities to empower a force structure to kill, but its potential is more far-reaching than just in waging kinetic operations. For example it provides for the efficient use and planning for medical requirements.[58] Counterinsurgency warfare is not about just killing. It is not fundamentally based on force-on-force engagements in which technical capabilities for kinetic operations are at the most optimal. This book examines and describes technology as an enabler of wide-reaching abilities, such as intelligence, mobility, and logistics, that are extremely important in counterinsurgency and even more so than in conventional engagements.

It is easy to focus on technology within the realm of force structures as it pertains to defeating the enemy force structure. The American mind-set and cultural attitudes often concentrate on technology far more readily than on the other capabilities that provide support to the military objective and hence the policy objective required in counterinsurgency warfare. The old adage that the pen is mightier than the sword still applies today and must be recognized as an incredibly important facet of warfare, particularly counterinsurgency warfare, where ideas, ideologies, beliefs, and identities are as important as military capability.

IV. The Multidisciplinary Nature of the Question

To address the research question and its different facets adequately and to fully explain the role of technology in U.S. counterinsurgency operations require different methodological approaches. Foremost the question must confront three separate issues.

To properly place the contextual question of technology in counterinsurgency operations, there has to be some overarching strategy for implementing the counterinsurgency practice. The surveyed writers on counterinsurgency have the unifying theme of protecting the populace in order to remove the support structure of the insurgents. This strategy not only makes sense but also is the likeliest to succeed within the limitations placed on U.S. military operations. To begin the process of answering the question, therefore, it is imperative that the reader understands that the goal of the strategy will always be to achieve dominance over the center of gravity—that is, the minds of the population—while it is also targeted by the insurgents. This assumption is touched upon continually in each chapter but is not explored in depth for two reasons: it is a commonly accepted counterinsurgency strategy, and, more important, this work is about the role of technology in U.S. counterinsurgency strategy, not the evolution of counterinsurgency strategy.

While the book continually refers to the need to protect the populace, the only methodological practice to support this claim is the self-evident nature of the idea as put forward by the surveyed authors on counterinsurgency. This claim, therefore, comes from a survey of twentieth-century texts.

With a strategic context for the conduct and practice of counterinsurgency operations in place, the second facet of the question that must be addressed is the tactical framework in which the strategy will be applied. While strategy is required, the actual implementation of the counterinsurgency strategy on the ground, at the tactical level, determines the feasibility of the strategy. In essence, there must be a harmony between the strategic, operational, and tactical levels of counterinsurgency practice. Without all levels working to a single goal, the net effect will be failure just as surely as with a lack of unity of command.

Methodologically the proper tactical role requires analyzing historical texts to identify a technical historical trend that fits into the strategic requirements of counterinsurgency practice. Without a historical reference no technical trend can be sufficiently identified to extrapolate into tactical doctrine. In the absence of such a trend, engineers and practitioners are left pondering individual widgets without context or tactical *and* strategic relevance. As such the operationally offensive, tactically defensive context is examined using historical texts of relevant battles to illustrate the effect of technology on the concept.

The third aspect of the research question that must be addressed is which technologies should be used and fostered and in what roles they should be used in U.S. counterinsurgency operations. Moreover, this facet of the research question must be considered within the context of both the strategic direction of the counterinsurgency campaign (winning hearts and minds) and the tactical (operationally offensive, tactically defensive) framework. No system should stand alone and be judged solely upon the merits of its technical military capability. A really fast tank may be pleasurable to drive and hard for the enemy to hit, but what is the tank's role and how does it fit in the strategic and tactical levels of counterinsurgency operations?

Methodologically determining which technologies and their roles are most appropriate required textual analysis, web resources, firsthand observation, and interviews. While the textual and web resources are fairly straightforward, the firsthand observations and interviews were not. Firsthand observations required conducting interviews with military and civilian personnel in Iraq while observing actual combat operations and civil affairs actions. While in Iraq and afterward, the interviewees were practitioners rather than academics. The reason for this choice is that technology must fill specific tactical needs, and practitioners with firsthand knowledge of real-world requirements can best address them.

In presenting the research question to be addressed throughout the book, the introduction lays the strategic foundation that to successfully wage counterinsurgency operations requires that the center of gravity must be in winning the hearts and minds of the population. Everything that follows must eventually be in harmony with this one overarching precept.

The first chapter examines the historical case studies that demonstrate the viability of the operationally offensive, tactically defensive concept. The concept is not applied to counterinsurgency operations; rather, its introduction and analysis are meant to show that a historical trend exists. The trend shows clearly that that with technology, through target acquisition and firepower, the number of individuals required to defend a chosen piece of real estate has continually decreased. The examples covered are conventional ones as no suitable examples can be gleaned from historic counterinsurgency operations.

Chapter 2 brings the context of U.S. counterinsurgency operations up to date. By examining the U.S. involvement in Iraq, a sense of current U.S. military capability and effectiveness can be understood in terms of the technology employed and of the doctrine being implemented. This chapter also provides the basis for many of the interviews in which the testimony of firsthand practitioners can be best contextualized.

The third chapter addresses the limits of the concept of counterinsurgency operations and current U.S. counterinsurgency operations. As the strategic underpinning of such operations is winning hearts and minds, it is imperative that this work recognizes even that particular strategic assumption has limitations. As such it is not a foregone conclusion that the operationally offensive, tactically defensive concept can be applied universally. In fact it will likely be problematic if the United States ever concludes that a single silver bullet could forever make counterinsurgency operations a forgone conclusion. This chapter attempts to curb the possible enthusiasm for the concept with a dose of reality.

Chapter 4 addresses technology itself and the role it plays and will continue to play in U.S. counterinsurgency operations. While this chapter is forward looking, it builds on the previous chapter's context: technology does not stand alone as a nebulous military capability without direction of purpose. Parts of this chapter are speculative; it cannot be avoided with a book of this type.

The next chapter addresses related contextual issues that will affect U.S. counterinsurgency operations. Having laid out the strategic, tactical, and technical framework for the role of technology in U.S. counterinsurgency operations, chapter 5 is meant to contextualize those operations. Without context this book cannot effectively address the real-world concerns of the role of technology in U.S. counterinsurgency operations. Just as technology itself is simply capability in a vacuum when it stands alone, so too is the idea of U.S. counterinsurgency operations when viewed without the political context in which they must exist.

Taken as a whole, this book provides the foundation to examine the question of the roles of technology in U.S. counterinsurgency operations and then addresses those roles and the context in which the operations take place. This book may be read as prescriptive, but the true intent is to dispassionately address the glaring gap in the literature of counterinsurgency operations in which the strategic relevance of technology is ignored.

1 The Operationally Offensive, Tactically Defensive Concept

> Dien Bien Phu . . . history doesn't always repeat itself. But this time
> it will. We won a military victory over the French, and we will win it
> over the Americans too. Yes . . . their Dien Bien Phu is still to come.
> GEN. VO NGUYEN GIAP

> Ha! There isn't anyone who is going to overrun my unit.
> CAPT. JOHN FRY

I. Foundation of the Concept

The *operationally offensive, tactically defensive concept* is a conventional
operational and tactical method that allows one party in hostilities to
choose the time and place of an engagement. The concept rests upon
the assumption that a tactically defensive stance is inherently stronger
than offensive operations in a force-on-force engagement.[1] The concept
blends the audacity and initiative of an offensive operation, choosing the
place and time of battle, while at the same time taking advantage of the
inherent strengths of a defensive mode of modern warfare.[2] It should also
be clear that as with all operations in warfare, any such operations take
place within a context of political objectives.

The elegance of an operationally offensive, tactically defensive concept
frequently looks more attractive on paper than it does on a given battlefield. Often each force is well aware of the strengths and capabilities of
the opposing force. Therefore, any force that chooses the place and time
of battle, prepares the battlefield with defensive works, and specifically
intends to bring its technological advantages into play will likely not be
attacked except as a last resort. There are numerous reasons why using
the operationally offensive, tactically defensive concept is inherently risky
and generally unlikely to succeed.

The nature of the concept requires that the force employing it risks encirclement and annihilation while forgoing the ability to operate offensively. While not guaranteed, often the force applying the operationally offensive, tactically defensive concept will be divided, thus leading to the inevitable, and potentially catastrophic, possibility of its defeat by superior numbers of an enemy force. Throughout history few generals have been able to divide their forces in the face of a superior enemy and bring engagements to successful conclusions as Robert E. Lee did with his masterful performance at the battle of Chancellorsville.[3] Historically the number of troops on opposing sides has been the primary criterion for determining which side has an advantage or is the superior force; however, such quantitative methods are not necessarily the best indicator of which side in a conflict is superior, especially in an operationally offensive, tactically defensive engagement.

The offensive side of the operation typically involves penetrating enemy territory or garrisoning disputed territory that is threatened by an enemy. Penetrating enemy territory provides the opportunity to exploit the enemy's rear, to interrupt lines of communication, and to conduct interdiction operations.[4] In such a scenario, having enemy forces in the rear area that create havoc and friction and participate in small-scale harassing attacks is often unacceptable. In this situation, forces must be mustered to attack and annihilate the pocket of resistance, which should be theoretically cut off from outside reinforcement. Such a deep penetration would allow the penetrating force to choose the point in space and time in which to force a battle on its own terms. Then the tactical advantage of the defense is determined by its technical capabilities, but it is likely to attract a high number of enemy forces. This important note needs to be understood: this concept provides for seizing ground and winning decisive engagements through tactical defensive superiority when the penetrating force is threatened. While the tactical advantage may be appealing, however, the stark reality is that the unit is surrounded and cut off in enemy territory without the hope of resupply and relief in a two-dimensional context, or a three-dimensional one, depending upon air defenses.

In the defensive role in disputed territory, the operationally offensive component is less important and sometimes not even a factor. This situation, of course, would depend on the nuances and the specifics of the battle in question and the disposition of the opposing forces in the area. However, the principle of strength in defense still remains technically driven and increases the likelihood of holding terrain. A classic example that illustrates the defensive advantage of technology is the 1879 battle of Rorke's Drift, which saw 139 superiorly disciplined British soldiers equipped with the breech-loading rifle hold off 4,000 Zulu warriors.[5]

The opportunities for employing the operationally offensive tactically, defensive concept are few in actual warfare. Its associated risks usually outweigh the possible benefits of the concept's employment. Consider President Dwight Eisenhower's appraisal of the French Union Army's position at Dien Bien Phu: "Finally they came along with this Dien Bien Phu plan. As a soldier, I was horror-stricken. I just said, 'My goodness, you don't pen troops in a fortress, and all history shows that they are just going to be cut to pieces.'"[6]

While the risk, militarily, may be manageable—as in the case of Operation Market Garden in World War II—the political cost associated with failure usually will be high as in the case of Dien Bien Phu.[7] However, there is an important caveat to the likelihood of success, and therefore the associated risks: the operationally offensive, tactically defensive concept's likelihood of success is directly related to the technical capabilities of the parties engaged in hostilities. The closer to parity the technical capabilities the less likely the concept will work without producing a catastrophic failure and/or high political cost such as the fate of the British First Airborne Division in Operation Market Garden and French Union forces at Dien Bien Phu.[8]

Technical asymmetry between two parties will obviously lead to differing military capabilities for each. As the nature of guerrilla warfare recognizes, the asymmetry between the two forces allows insurgents to take advantage of any specific weaknesses of large conventional forces, such as a plethora of targets and opportunities within the lines of communication and logistics trail. As Samuel B. Griffith noted: "In every apparent disadvantage, some advantage is to be found. The converse is equally true: In each apparent advantage lies the seed of disadvantage. The Yin is not wholly Yin, nor the Yang wholly Yang. It is the wise general, said the ancient Chinese military philosopher Sun Tzu, who is able to recognize this fact and to turn it to good account."[9]

Conversely in such a scenario, the conventional forces can establish heavily fortified defensive outposts in enemy terrain that are extremely resistant to being attacked and overrun by smaller forces. In that instance, the normal mode of operation for the guerrilla or insurgent is simply to fade away and deny battle to stronger conventional forces while looking for targets of opportunity.[10] The important point here is that technical capability allows for enhanced defense through firepower and protection, thus allowing for an attrition of enemy forces; but an insurgent force is unlikely to enter into such a disadvantageous, attritional style of confrontation.

II. Gettysburg

An early example in U.S. history of the concept's possible application was prior to the battle of Gettysburg in the American Civil War. General Lee was convinced that an offensive operation into the North would have a strategic impact on the course of the war. One of his most able corps commanders, Lt. Gen. James Longstreet, felt that the operational concept of striking in the North to seek a decisive engagement could be better achieved through maneuvering into a defensive position rather than by attacking or seeking engagement as would happen at Gettysburg.[11]

It is often said that the technology of the Civil War outpaced the military tactics and the thinking of the commanders at that time. The advent of the minié ball and the subsequent increased range in killing power of individual weapons, combined with tactics that did not take advantage of cover and concealment, led to horrendous slaughter on the battlefield.[12] As the technical tools of warfare began to change during the Industrial Revolution, the subsequent impact of those technical innovations was not always appreciated on the battlefield. In the 1860s soldiers fighting from defensive works and dug-in positions clearly had an advantage. In Longstreet's view, it was the spade for digging and the rifle and cannon for killing that provided his soldiers with an inherent advantage when in the defensive posture. This judgment, of course, was a change from the historical perspective that the attacking force had the advantage.

Conventional wisdom regarding tactical engagements stipulates that an effective offensive force controls the initiative and tempo of an engagement. As Gen. George Patton said: "Wars are not won by defensive tactics. You keep moving and the enemy cannot hit you. When you dig a foxhole, you dig your grave."[13]

A defensive force may only react and thus is at a disadvantage as it will be unlikely able to maneuver and regain an advantage on the field. This thinking, while generally accurate, has led to a culture and belief that when in doubt the best thing to do is to attack. For General Lee the notion of invading the North, moving to a set piece of real estate to construct defensive works and await the enemy, would give away his ability to maneuver; it would surrender the initiative to Union commanders who potentially could surround and starve his Army of Northern Virginia. That course of action was unthinkable for such a general as Robert E. Lee, who was a master at maneuver and so confident and knowledgeable that he could consistently divide his forces in the face of a numerically superior enemy and win engagement after engagement. It is a question for history whether Lee understood the technological implications of the minié ball or whether he was confident in the conventional wisdom of maintaining offensive operations when possible. The question is also academic as the course of history unfolded with the fateful battle of Gettysburg.

If Longstreet's counsel had been heeded, the Army of Northern Virginia would have maneuvered itself to a location of defensive advantage between Washington DC and the Army of the Potomac, specifically its supplies at the Westminster railhead.[14] It would have been a classic example of the operationally offensive, tactically defensive concept put into action with all of the associated risks. It was Longstreet's view that the Army of the Potomac, with its lines of communication to Washington DC cut, would be forced to assault the prepared defensive positions of the Army of Northern Virginia. As he wrote: "I suggested that, after piercing Pennsylvania and menacing Washington, we should choose a strong position, and force the federals to attack us, observing that the popular clamor throughout the North would speedily force the Federal general to attempt to drive us out. I recalled to [Lee] the battle of Fredericksburg as an instance of a defensive battle, when, with a few thousand men, we hurled the whole Federal army back, crippling and demoralizing it, with trifling loss to our own troops."[15] Fatefully General Lee chose not to follow Longstreet's advice, a decision that resulted in the battle of Gettysburg and the end of any real hope for a Southern victory in the American Civil War.

With Gettysburg, Longstreet's desire to move operationally into Northern territory showed audacity. It also showed that Longstreet, as a general, was capable of understanding the political context in which the war was being waged. He was keenly aware of how the North's political reality would force a Northern attack on Southern troops entering Northern soil regardless of the tactical considerations of the field of battle chosen by Southern forces. Longstreet recognized the synergy between the number of soldiers defending a given objective, their ability to prepare the objective to enhance its defensive value, and the technology available through spade and minié ball to take advantage of the defensive posture. This combination of technology, tactics, operational concepts, political narrative, and intuition would eventually lead to a coherent understanding of the role of technology in counterinsurgency operations in modern times while using the operationally offensive, tactically defensive concept.

A point that should not be missed from this Gettysburg example is that Longstreet understood that the Army of the Potomac would be forced to attack the Army of Northern Virginia regardless of the terrain and tactical advantages held by the Southerners. The true genius of Longstreet's plan, and the potential Achilles' heel for the North, was that political realities would have forced the North to offer battle at a disadvantage. Longstreet understood warfare within the Clausewitzian terms of it being an extension of politics. Any force at anytime throughout history that feels it has no choice but to attack when in a disadvantaged situation is in a very unenviable position. Theoretically putting an enemy in this position is not hard; however, it is incredibly difficult to do so without incurring more risk than is acceptable.

In a hypothetical example, which would never have been seriously considered, the Allies could have dropped multiple airborne divisions behind German lines during World War II to march on Berlin. Such an operation would have been doomed to failure and would have served no viable strategic end. Without any doubt, the airborne units would have been surrounded and defeated. Equally without doubt, the Germans would have felt compelled to attack and annihilate the airborne units just as they did during Operation Market Garden. The skill is not in forcing an enemy to attack a defended position for defensive advantage but in choosing a position that not only the enemy must attack but also offers both defensive advantage and strategic advantage, thus representing low risk and high benefits. Longstreet's concept is a classic example that fits the criteria for an operationally offensive, tactically defensive campaign, but the opportunity it presented was missed. The level of technology available to the North and the South was similar at the tactical level even though the South relied on imports. Although there was no symmetry in production capability or in manpower to utilize the production capability that gave an overwhelming advantage to the North, a Northern spade could dig as much dirt as a Southern spade and shots fired by Northerners were as lethal as shots fired by Southerners.[16] So as noted before, the chance of success of Longstreet's plan was not going to be dictated by an asymmetrical technical advantage of Southern forces to enhance a defensive posture and inflict tremendous casualties on Northern forces through superior firepower. This lack of technical asymmetry naturally led to a requirement for a large number of forces to be used in battle. The simple arithmetic of attrition warfare, as with the French strategy of *grignotage* (nibbling) in World War I, is sufficient to understand the interaction between Northern and Southern forces during the American Civil War.[17] The real technical advantage that a defender had during this period was the spade. By reducing his body's exposure to enemy fire, the defender had a far greater chance of survival than did troops maneuvering in the open.

An important point here is that technology, as an aid to defensively postured troops, has a direct correlation to the size of the force needed to defend a territory against an attacking enemy. Clearly the number of troops required to defend a given point during the American Civil War depended on the number of troops attacking them. So while the defender did have an advantage through the digging power of the spade, it was a foregone conclusion that given a large enough numerical advantage, troops could always overrun defended positions because the level of defensive technology was not yet very advanced.[18] As noted earlier, the operationally offensive, tactically defensive concept provides the ability to attrit enemy forces and hold ground. Given enough firepower and a sufficiently large body of troops, Northern forces could overrun dug-in Southern forces, but they would pay a heavy price in doing so.

The Longstreet example from Gettysburg is so poignant because in addition to showing the potential of the operationally offensive concept with the likely benefits of the tactically defensive components, it demonstrates a connection and synergy between the tactical level, the operational level, and the strategic level with its intended political consequences. Longstreet clearly understood the tactical advantage and wanted to capitalize on that tactical advantage and the operational concept. He went even further in understanding that the concept could use attrition and the retention of ground taken to proper political effect with potential strategic consequences.

III. Artillery and Machine Guns

The most fundamental components of a defensive position consist of firepower and protection for the soldiers in the position. Obviously protection reduces the vulnerability of defending soldiers to incoming enemy fire whether it is from advancing soldiers or standoff weapons such as cannon artillery.[19] In the most basic explanation, a soldier defending a position typically will expose the minimum amount of body mass necessary to defend the position. Ideally the exposure will be limited to the head and shoulder only insofar as is required for the soldier to discharge a weapon in defending the position, and preferably the head will be protected by some sort of body armor. Such a limited exposed area becomes very difficult to hit at range by opposing infantrymen, especially as the attacking force attempts to advance. Likewise, the effects of incoming artillery can be seriously reduced, and often negated, by the construction of bunkers and solid overhead cover for troops defending the position.[20]

In the Gettysburg example, the complexities of offense versus defense are fairly simplistic. Soldiers, whether from the North or the South, could prepare defensive positions simply by digging into the soil and reinforcing trenches with materials such as wood. The most likely threat the defenders faced was from massed infantry assault supported by cannons, which could fire both kinetic and explosive rounds.[21] Neither the shrapnel shell of the day nor direct-fired solid shot, however, would be likely to penetrate an even moderately well-built bunker that was not directly facing the enemy assault. Therefore, within the technological context of the American Civil War, infantry dug in. Using the spade as their technical tool, the men could stand a reasonable chance of defending any position against a relatively equal number of advancing enemy forces.

The defense had become so strong that from 1860 through World War II the soldiers' ability to dig and build reinforced bunkers and to deliver devastating rifle fire, and later machine-gun fire, with pinpoint accuracy made assaulting well-prepared infantry positions an unsound tactic.[22] Naturally the evolution of military tactics dictated a need, or at least an effort, to overcome the strength of the defensive posture. There were, of course, periods in which attack was seen to be supreme in order to maintain initiative, to control tempo, and to retain the ability to maneuver. However, even a cursory glance at engagements in which defending soldiers were willing to fight and die shows a strong and disproportionate argument in favor of defense.

During World War I, trench warfare became the norm as infantry dug in along long lines of fortifications. Throughout the war, such as the battle for Verdun, large numbers of infantry often attempted to assault and break through the defensive trenches of the opposing side.[23] In the meat grinder of the western front, such attacks were often successful in reaching and occupying the first line of enemy trenches, but the inevitable counterattack and massive artillery barrages would soon drive them back to the original point of departure.[24] Given the technical capabilities of the militaries fighting on the western front during World War I, the men simply did not have the ability to maneuver quickly enough to exploit any breakthrough in the enemy lines to move into the enemy's rear and create havoc in their lines of communications.[25]

Defense during World War I had become so powerful based solely on technology. The ability of a soldier to dig into the ground or to build defensive positions was not new to the twentieth century. The power of defense to give soldiers an asymmetric advantage over an attacking force was also not new. What was new during World War I were the technologies used to support the defensive posture of infantry in the field; they dramatically improved the firepower and killing effect of individual infantry weapons and indirect artillery fire. A musket used in the American War of Independence typically had a smoothbore and an effective range of no more than 120 yards, but it could be quasi-accurately aimed out to 100 yards. By the Civil War, the minié ball had effectively extended the range to 500 yards and increased the accuracy and chances of hitting a man-size target.[26] By World War I, the standard U.S. infantry weapon was the 1903 Springfield rifle with a .30-06 round capable of consistently hitting targets the size of a dinner plate at 500 yards and, in the hands of a truly talented shooter, engaging targets at 1,000 yards or more.[27]

In a crude mathematical formula, the rate of fire of a musket is about two to three rounds a minute, depending on the proficiency of the shooter, with a maximum effective range of 120 yards.[28] Assuming that an attacking wave of soldiers can cover two yards a second, a defender using an eighteenth- or nineteenth-century smoothbore muzzle-loading weapon will only be able to fire two to three rounds at the attacking forces, with fairly low probabilities of actually hitting an enemy soldier with each round fired, before the attackers can close with the defenders. In essence if an attacking force is able to muster three times the number of a defending force, as during the American Civil War, the attacking force would almost be assured of successfully carrying through the attack and seizing the objective. By World War I, this crude mathematical formula was no longer an effective measure of the strength of a defending force as rates of fire and the range of weapons had increased dramatically.

While individual rounds fired by infantrymen in a defensive position had a devastating effect when fired from modern rifled firearms, a machine gun, such as the Maxim machine gun of World War I, had an effective rate of fire of six hundred rounds per minute. The gun put an even deadlier and more accurate amount of firepower in the hands of soldiers in defensive positions who already had an advantage.[29] A graphic example is the September 25, 1915, attack of two British divisions on the German lines at Loos. Of the 10,000 British soldiers who attempted to assault the German position, 385 officers and 7,861 enlisted men became casualties with few German losses.[30] The majority of the casualties were inflicted starting at about fifteen hundred yards from the German lines as the German machine guns opened up on the advancing British troops. Technology had become such that defending soldiers were able to deliver a high enough volume of fire, accurately, to make massed infantry assaults an almost obsolete tactic that was likely to accomplish nothing.

The most important point to note about the effect of the machine gun, as with any ranged weapon, is its ability to successfully trade time for space in favor of the defending force. The weapon does not simply kill enemy soldiers; rather, it engages enemy soldiers at its maximum effective range and slows their advance. In essence it forces the enemy to move through the space where they cannot successfully engage the defenders over the longest period of time possible. By trading time for space, the defenders create casualties in two dimensions for every unit of space that the attacking force crosses. As the attacking force moves closer to the objective, it takes casualties from the defenders' fire, but the attacking force must also be very aware that, with every step, it pays for moving toward the objective. If the attackers fail, they will have to pay for the space when moving away from the objective. The psychological effect is that a soldier, unless absolutely convinced of success, will want to progress as short a distance as possible before turning back. To make matters worse, a competent defensive force will construct obstacles to slow the advance of the attacking infantry; barbwire and stakes were effective defensive tools on the western front.

The strength of the defense and the defenders' ability to trade time for space against an attacking force are directly linked to the technological capability of the defending force. Through the Gettysburg example, clearly the defending forces, be they Northern or Southern, had an advantage over their adversaries based on the spade's providing the infantry with the ability to dig fixed positions. However, the technology of their firearms (rate of fire) meant that they were not impervious to infantry assaults by an adversary. By World War I, the technology associated with the weapons used by defending infantry, from fixed defensive positions that had been well prepared, made the defenders almost impervious to human wave attacks by opposing infantry.[31] The rates of fire and the accuracy of individual and crew-served weapons specific to the infantry had made the price of attack so high as to make a complete breakthrough in enemy positions almost impossible without a faster means of maneuver than moving on foot.

The potential of the combustion engine was just starting to emerge during World War I. Neither motorized ground transport in the form of mechanically reliable armor nor air assets, such as heavy bombers, were yet a realistic option as military tools on a strategic level. The technology was nascent and clearly had the potential to build into effective tools for warfare, such as early attempts by the British to develop and use tanks on the western front during World War I, but it was not capable of breaking the stalemate of trench warfare at that time.[32] Fundamentally the problem faced by attacking forces was one of mobility. Theoretically enough artillery firepower, to compensate for inaccuracy, would be brought to bear on a point of the enemy line to effect a breakthrough into the enemy's rear area, where troops would pour in and exploit the success to take ground, flank enemy positions, and disrupt lines of communication.[33] Unfortunately for the soldiers on the western front, there were no large-scale maneuver answers to the horrors of trench warfare. World War I progressed with the continual effort to smash holes in the enemy's lines without a proper understanding of the killing power of modern weapons.[34] When infantry was used to directly assault enemy trench systems but had almost zero success at breaking through, massed artillery was used in an attempt to annihilate a section of enemy trenches.

Artillery represented the only practical way to deliver explosive power to the enemy's defensive position and to reinstate mobility by achieving a breakthrough.[35] The explosive power required to clear enemy trenches theoretically could be achieved through massive artillery bombardment, which churned the earth into overlapping shell craters; however, extremely well-fortified and deeply dug-in bunkers with large amounts of overhead cover are capable of withstanding almost any amount of artillery bombardment.[36] The bombardment may be very disconcerting for soldiers inside the bunkers, pounded continuously and hit by frequent shock waves, but defending soldiers in and around the area of bombardment are unlikely to be completely annihilated.[37] So while artillery can deliver large amounts of ordnance onto defensive positions, its effectiveness is hampered by well-developed and fortified defensive positions designed to protect the infantry.[38] The artillery's effectiveness in such a role is also hampered by the likelihood of reserve units outside the area of bombardment being able to move easily into the area once the bombardment has ended.

As artillery and preparation fires lifted from a section of enemy lines before the assault by friendly infantry forces, they also signaled to reserve forces that it was time to move up and reinforce the depleted section of the defending line.[39] Assaulting infantry formations could often reach the enemy trench line and even take the first several lines of trenches that had been devastated by the artillery, but they could not take in-depth defensive positions.[40] Only heavy bombardment and inadequate defensive preparations would allow assaulting infantry to take a trench line successfully by the end of the war, but they would still face the remnants of the defenders and those reinforcements moving up to reoccupy the lost positions.[41] The assaulting infantrymen would also face artillery fire both to slow their advance and to harass their withdrawal.

Early World War I battles showed artillery was far more effective for the defender than it was for the attacker. Consider J. F. C. Fuller's analysis: "The . . . disproportion between losses and gains. For example, in the battle of Third Artois–Loos, the French and British respectively lost 48,200 and 48,267 men, and in the Second Battle of Champagne the French losses were 143,567. In both, no more than the German front line system of trenches, in places some 3,000 yards deep, was captured."[42] Artillery works best when used against targets in the open, or when moving in the open, for it allows artillery to maximize the killing radius of outgoing rounds. While its effect on troops in the open is undeniably devastating, the effects of artillery on well-dug-in troops with good overhead cover in a defensive position are minimal unless truly vast quantities of shells are used to saturate relatively small areas of defensive fortified structures. It should be noted that artillery can be used very effectively against defensive positions if the defenders, either through a lack of time, materials, or discipline, have not properly defended the objective by constructing bunkers and overhead cover.[43] In such circumstances, artillery will have an inevitable and costly impact on the defending forces.

It would appear that by World War I the operationally offensive, tactically defensive concept would have worked well on technological grounds. Defense had evolved more lethal technical means of offering protection at longer ranges to allow a more robust trading of time for space. In short the defensive tools were available to make defense a potent capability in comparison with the past. This point, however, is somewhat misleading as the real answer is counterintuitive. In fact mobility to facilitate the operationally offensive side of the concept was utterly lacking. Infantry units were mobile during World War I, except obviously on the western front, which was a single continuous front from the English Channel to the Swiss Alps; but the movement of their supplies—specifically the attendant artillery to provide the firepower required to effectively trade time for space in a defensive effort—was not practical.[44]

World War I marked the advent of two important facets of the operationally offensive, tactically defensive concept. First the amount of firepower available to infantry units heralded a new age of infantry combat whereby it was no longer reasonable to expect assaulting infantry to be able to overrun defending infantry without vastly disproportionate numbers in favor of the attacker. Infantry battles would never again resemble those of the Napoleonic era. Second technology had provided the means for the defense to trade time for space in a truly effective manner. As shown by the musket example earlier, a couple of rounds fired over a distance of 120 yards, while beneficial for the defense, is not an effective way to trade time for space given the average pace at which a man moves. The technology of the machine gun, indirect artillery fire, barbwire, and individual infantry weapons extended the killing zone around defensive positions far enough so that the trading of time for space became a maturing asset to the defense. What was missing for the operationally offensive, tactically defensive concept to really come into its own was a revolution in mobility, and it was about to take place with the arrival of the airplane.

IV. Dien Bien Phu

In November 1953 French Union forces in Indochina began to implement an operationally offensive, tactically defensive operation in a river valley at Dien Bien Phu.[45] The operation was a daring attempt to cut off

east–west travel by the Vietminh forces between the Red River Delta and the French Laotian west, as well as to provide a stronger defensive position for French Union forces in the highland area. Dien Bien Phu was also surrounded by rich agricultural lands and was therefore of strategic value to the Vietminh.[46] As the French were about to find out, operationally offensive, tactically defensive campaigns are extremely risky affairs, where technical capabilities can make the difference between success and failure.

The French counterinsurgency campaign against the Vietminh in Indochina was a prelude to the U.S. intervention in the same area a decade later. The French attitude at the time followed much of the conventional thought on warfare that was focused around an effort to close with and kill or capture enemy military formations in order to defeat the insurgent threat.[47] Although these operations took place before a mature appreciation of modern counterinsurgency techniques had become widespread, insurgent techniques as developed by Mao Zedong were fairly well understood and served as a practical example of success in China, Vietnam's northern neighbor. In fact the Vietnamese would directly implement Mao's three stages of revolutionary warfare and attempt to drive the French colonial administration out of Indochina.[48] An important note here is that it was not simply a communist effort to remove French colonial control of Indochina; for many Vietnamese, nationalistic fervor was a stronger driving factor.[49] Originally most Vietnamese identified with nationalism rather than the economic and political ideology of communism.[50]

The disadvantages of a conventional mind-set applied to counterinsurgency situations will directly hamper any chance of success for counterinsurgency forces and the long-term struggle for the allegiance of the populace. Perhaps the French can be forgiven for their overly conventional approach given the recent events in World War II; however, the French failed spectacularly to appreciate or integrate sound ideas of counterinsurgency doctrine and practice in Indochina. Later experienced French officers such as David Galula would outline them.

It is also important to note that by 1953 the Vietminh had organized themselves into conventional units. These units had regularly engaged in conventional warfare against French units both to the north and west of the Red River Delta. In fact General Giap had prematurely attempted an all-out conventional assault on French Union forces in the Red River Delta in the 1951 battle of Vinh Yên and suffered a crushing defeat due to the firepower and discipline of the defending French Union forces.[51] It must be emphasized that by 1953 the Vietminh forces consisted of large bodies of well-trained men with sufficient logistical support to be considered a serious threat to French control of northern Indochina.[52]

The French would have benefited from sound counterinsurgency theory and practice earlier in the struggle for Indochina rather than later. The material resources and manpower required for successful counterinsurgency operations tend to be much lower at the inception of the insurgency rather than later.[53] The difficulty, of course, is in realizing through solid intelligence that an insurgency is in its nascent stages. The French were so far behind in their counterinsurgency campaign in Indochina that they already had a serious problem before they came to grips with the fact that they had any problem.

As the threat of the communist insurgency in Indochina and the extent of the support that communist China was providing the Vietminh became clear, the French began a strategic effort to clear insurgent elements from Indochina through engaging Vietminh forces and destroying their rear base areas to reduce their capacity for continuing operations. This plan shows that the French fundamentally failed to understand the center of gravity in the counterinsurgency effort they were undertaking. This flawed thinking guided their actions, which, in hindsight, can clearly be seen as attempts to crush the insurgency by crushing its forces.

Even if the French had understood that the center of gravity was in the minds of the people rather than in the forces fielded by the Vietminh, they would still have found themselves in an extremely difficult situation as they were unable to protect all of the people by separating them from the insurgents.[54] While the French were aware of the role and importance of the people in the conflict and did hold urban centers, clearly they were never able to tailor their strategy to effectively address the issue.[55] The center of gravity in counterinsurgency—or in other words, the source of power that provides moral or physical strength, freedom of action, or the will to act for the adversary—is in the minds of the people and not in the insurgent formations until much later.[56]

As the Vietminh slowly nibbled away around the periphery of French Union forces and French Union forces swept through the jungle and attempted to draw the Vietminh into a decisive engagement, the French were slowly bleeding themselves dry. Gen. Henri Navarre described Dien Bien Phu as a mooring point for operations in northern Indochina, and Maj. Gen. René Cogny called it a resupply point for tribal guerrillas. Cogny later said, "I suggested the occupation of Dien Bien Phu to install there a simple mooring point for our military and political activities in Northwestern Tonkin. We thus would derive an advantage from the hostility of the T'ai mountaineers against the Viet-Minh from the plains who seek to subject them to their yoke. Unfortunately, the capital of Lai Chau cannot even be defended against moderate attack."[57]

In essence Cogny wanted to establish a defensive post because the current center of activity, Lai Châu, was not defensible; he proposed an operationally offensive move into the Dien Bien Phu area in order to establish a defensive position. While the two men differed on the actual purpose of Dien Bien Phu, both would agree that it was a remote location in which a forward base capable of defending itself would be constructed. Even though Dien Bien Phu was to be a mooring point for operations in northern Indochina, its remoteness and difficulty of access, realistically only by air, made it very unlikely to be a successful base for large-scale offensive operations in the area. General Navarre's conception of Dien Bien Phu as being a mooring point for offensive operations simply cannot be reconciled with the French resources in Indochina at the time. Rather, General Cogny's conception of Dien Bien Phu as a defensive and resupply hub for indigenous tribal guerrillas seems much more likely, making Dien Bien Phu a classic example of the operationally offensive, tactically defensive operation.

As history has shown, Dien Bien Phu was a catastrophe for French Union forces through incorrect assumptions about artillery support, ineffective preparation, and, most important, a lack of firepower. The operationally offensive, tactically defensive concept that would validate such an operation as the occupation and fortification of Dien Bien Phu needed to rely on firepower as its primary mode of defense. While, as noted previously, innovations in infantry weapons made the infantry extremely lethal and artillery support was exceptionally effective in the defense of infantry against attacking forces in the open, the defenders had to be well dug in and have the technical tools of firepower at their disposal. In 1953 the technology of firepower to provide an almost impenetrable defense for the infantry was not available to French Union forces in Indochina.

Never exceeding thirteen thousand to fifteen thousand men at any one moment, the garrison at Dien Bien Phu had around twelve thousand men in mid-March, and these men were slowly being depleted as the fifty-six-day siege progressed.[58] The French forces were able to hold off the Vietminh siege force of almost fifty thousand soldiers with another fifty-five thousand support troops until May 7, 1954.[59] So while historically, such as in the American Civil War, a 3:1 advantage for the attackers over the defenders could provide a reasonable assurance that an assault would take a defended objective, by 1953 a 3:1 advantage was nowhere near enough to overrun the objective. It should also be noted that not all of the thirteen-thousand-plus men defending Dien Bien Phu were actual combat troops and that the defenders had only a minimal close air support and resupply yet they still inflicted twenty-three thousand casualties on the Vietminh.[60]

The telling story of the siege was the French forces' unfounded assumptions that the Vietnamese forces would be unable to move up to two hundred heavy artillery pieces cross-country from the Red River Delta area to the area around Dien Bien Phu.[61] The French artillery commander at Dien Bien Phu, Col. Charles Piroth, in fact, stated to General Navarre: "Mon General, no Viet-Minh cannon will be able to fire three rounds before being destroyed by my artillery."[62] Piroth was probably making the assumption that massed enemy artillery would not be available to Vietminh forces. The fact was that the Vietminh not only had the artillery dragged through the jungles to Dien Bien Phu but also managed to bring in copious amounts of antiaircraft artillery to suppress effective close air support and resupply to the French garrison.

In effect the French were undone due to faulty assumptions and their inability to employ effective counter-battery fire or air-delivered firepower to silence the Vietminh artillery that hammered their position. Because they hastily built the fortifications, providing proper overhead cover of bunkers was not possible. They had to fly in construction materials for the fortifications as there were not enough local materials to construct them adequately.[63] Also the garrison was located at a low point in the valley, which tended to flood, ruling out the possibility of deeply buried fortifications capable of withstanding bombardment by enemy artillery. In short Dien Bien Phu was a battle that resembled World War I on the ground (with the exception of the French having some tanks) without heeding the lessons of World War I.[64]

The fate of Dien Bien Phu should not be read as a failure or lack of viability of an operationally offensive, tactically defensive concept in warfare. As it turned out, the Vietminh had badly miscalculated and discounted the effectiveness of airpower. That the French Union forces did not have an adequate air capacity available to meet their needs in Indochina ignores the fact that the United States could have intervened with massive airpower to save the garrison but chose not to do so.[65] In fact Gen. Georges Catroux, who chaired the French government's investigating commission after the battle, said the "sole chance of salvation of the heroic garrison rested on a massive intervention of a fleet of American bombers based both on aircraft carriers of the United States Navy and in the Philippines. Undertaken by some three to four hundred heavy aircraft, that operation would have, in the view of the experts, smashed the Viet Minh [siege] organizations and would doubtlessly have reversed the course of events."[66]

The French Union forces' failure in the battle of Dien Bien Phu was both tactical and strategic. Constructing poor fortifications, overestimating the capability of artillery, and lacking both an appreciation of the environment and the appropriate technical tools all led to a tactical defeat of French forces at Dien Bien Phu and the ultimate defeat of French Union forces in Indochina. While two hundred to three hundred American aircraft would have made the tactical difference, they could not have made a strategic difference.[67]

The operation, however, did see some tactical success. Most notably for almost six months, French Union forces were able to tie down five communist divisions, which formed 60 percent of the total Vietminh main battle force and almost 20 percent of all their military manpower, including their guerrilla forces.[68] It was a fairly major achievement for only 3.3 percent of the French Union forces and shows that by 1953 the technology involved in providing artillery fire to support infantry units together with the improved quality of infantry weapons favored the defending forces even more strongly.[69]

The strategic failure of the French Union forces in Indochina was deep, and it should be noted that they conducted Dien Bien Phu at the operational and tactical level mostly cut off from what should have been synergy at the strategic level. As Bernard Fall stated, "Air power on a more massive scale than has been available could not have changed the outcome of the Indochina war, but it would have saved Dien Bien Phu."[70] So it could reasonably be concluded that Dien Bien Phu, with the increased technological capabilities of robust and high-altitude close air support, could have been an operational and tactical success, validating the operationally offensive, tactically defensive concept, but it would have failed to have a strategic impact on the true center of gravity of a counterinsurgency operation—that is, the mind of the populace and, thereby, the will of the insurgent.

Dien Bien Phu is an unfortunate example of the operationally offensive, tactically defensive concept being technically viable but under resourced.[71] As the Gettysburg example demonstrates, the concept has the potential to be used tactically and operationally and for strategic effect. It also shows that a knowledgeable general, Longstreet, did understand that the technology available at the time strengthened the concept and its military utility, even though that technology consisted of little more than the newly shaped rifle round and the spade for digging. The technical achievements of World War I, such as the development of accurate indirect supporting fires and high-velocity rifle rounds, provided the individual infantry unit with the particular firepower available to withstand historically overwhelming assaults. Dien Bien Phu represents an almost successful attempt to blend the technical tools of air mobility, close air support, and supporting artillery fire into a viable operational and tactical concept. Fall was correct that by 1953, close air support by heavy aircraft was a viable mode of operation for defending a beleaguered garrison; but while the technology was available, the resources were not.

The strategic failure of French Union forces in Indochina generally, and Dien Bien Phu specifically, was a failure to understand how to conduct counterinsurgency operations properly in order to effectively combat revolutionary warfare.[72] The French could not have been expected to use the technology and doctrine for strategic effect without both an understanding of where the enemy's center of gravity truly was and the willingness to act upon that knowledge. Many countries would not learn this lesson immediately. It does illustrate that while the operationally offensive, tactically defensive concept has military utility and the potential to be an effective force multiplier, no matter how much technology is applied, a misguided strategy that does not take into account sound practices and doctrine is bound to fail.

The French did realize to some degree the importance of the people to the communist cause and that the technology and killing power, though not always at their disposal, were robust. It is likely that putting these two facets together to protect all population centers would not have been possible without massive reinforcements, which the French did not receive. Without weapons and sensor technology, which did not really start to develop until the 1960s, the French did not have a chance of strategic success with the number of soldiers available to them in Indochina.[73]

V. Airpower

In Dien Bien Phu French Union forces relied heavily on airpower at the operational and tactical level to hold their position in the river valley. Airpower represented the supply lifeline for the besieged soldiers, as well as fire support through close air support.[74] Airpower also represented a third aspect, or interdiction, which was applied in Dien Bien Phu through an attempt at aerial interdiction of enemy supplies moving toward the besieging forces of the Vietminh. By 1953 airpower had the technical potential to provide outstanding results in resupply and close air support, but the ability for airpower to interdict enemy lines of communication was extremely limited and would continue to be so in the future.[75]

Aerial resupply of troops, especially those besieged in a rather small pocket, is a problematic process that endangers aircraft and aircrews and therefore reduces the likelihood of effective resupply for the men on the ground. A combination of factors, such as weather, enemy antiaircraft fire, and runway unavailability, can complicate any effort to resupply. In the case of Dien Bien Phu, the weather was a frustrating factor for French Union forces as often visibility was low and precipitation high. Aircraft also were almost continuously risking high volumes of antiaircraft fire from Vietminh positions and eventually were not even able to land on the runway. They had to resort to parachuting supplies, which often landed on Vietminh positions.[76]

Dien Bien Phu was an audacious move by French Union forces, but it can probably be seen as foolhardy or perhaps overreaching as the French Union forces simply did not have the air assets in place in Indochina to provide effective support to the beleaguered force. As noted, French Union forces might have won the actual battle of Dien Bien Phu with the intervention of American airpower to deliver large amounts of ordnance in the area around the besieged garrison. The technology for close air support, in fact, did exist in 1953, but the French could not deploy the aircraft required for an effective defense of a place like Dien Bien Phu. Certainly the ability to supply by air thirteen-thousand-plus men with basic essentials was well within the technological scope of aircraft in the 1950s but not at Dien Bien Phu as surrounding Vietminh forces used their unexpected antiaircraft artillery and posed a regular artillery threat.[77]

High-altitude carpet bombing might have been effective at Dien Bien Phu for breaking the besieging force and thereby allowing French Union forces to move from their poor fortifications into the field to close with and destroy the enemy artillery that pounded their defensive position. It would have been essential for any effective air-delivered ordnance to have come initially from aircraft at greater than nine thousand feet above ground level in order to minimize the risk of damage coming from the 37mm antiaircraft fire from Vietminh positions.[78] And again, there is nothing to indicate that in 1953 the technology was not available for aircraft to fill this role. In fact the U.S. Air Force B-29s flying in the Korean conflict at the time had exactly that capability. For example, Lt. Gen. George Stratemeyer considered the following solution to the antiaircraft problems around Pyongyang: "I proposed to direct a major B-29 strike against several of the most important remaining military targets in the Pyongyang area. Selected targets include several military barracks and training areas, warehouse, storage areas, and marshalling yards. Plans call for the use of 100 B-29s on this mission in order to saturate the antiaircraft defenses and insure the elimination of these important targets in one strike."[79]

This passage clearly shows the potential for high-altitude strikes from large numbers of aircraft in the East Asian theater. The French Union forces would obviously have benefited greatly from such air assets. Massive strikes by heavy aircraft to help break the stranglehold of the besieging force on the garrison at Dien Bien Phu would have provided the space for the defenders to move out to expand the perimeter and assault enemy artillery on the surrounding hills, thus allowing the effective aerial resupply of the defensive position. Such strikes would also have reduced or eliminated the antiaircraft fire that prevented the low-altitude close air support for the French Union forces. The ability of aircraft to provide low-altitude close air support was almost zero, and many pilots described the flak around Dien Bien Phu as being "as dense as anything allied pilots had encountered over the Ruhr during World War II."[80]

The French Union forces were unable to interdict the Vietminh supplies moving toward Dien Bien Phu to any meaningful extent owing to the variety of resupply routes taken by the coolies pushing bicycles carrying up to 440 pounds.[81] With tens of thousands of such individuals moving supplies, items ranging from rice to artillery, through dense overhead cover, French Union aircraft had little opportunity to spot or engage the supply line. Interdiction of enemy supply routes is almost impossible to carry out effectively when it is not clear where the supply routes are and when the supplies are moving. The United States would subsequently find it difficult as well.[82]

Aerial interdiction of supplies is an objective for which the technology of sensing capabilities was not advanced enough by 1953 to be effective in supporting a unit participating in an operationally offensive, tactically defensive operation. With supplies not moving along well-defined road networks, easily identifiable railways, and chokepoints such as bridges, the problem of finding and accurately targeting the individuals and equipment involved in the resupply of enemy positions and forces becomes almost impossible. In 1953 while the U.S. Air Force had the capability of delivering massive firepower from high-altitude aircraft, it failed just as spectacularly in interdiction efforts in the Korean conflict. Its Operation Strangle used hundreds of heavy and medium bombers in a yearlong effort of round-the-clock bombing to interdict the supply lines for Red Chinese and North Korean forces without success.[83]

Insights gleaned in the use of airpower and operationally offensive, tactically defensive operations are obviously not forthcoming from the American Civil War, nor from World War I except in a limited way, other than to suggest that aerial observation from aerial platforms provided intelligence.[84] However, by the time of the French defeat at Dien Bien Phu in 1953, the technology behind airpower to support an operationally offensive, tactically defensive operation was starting to evolve into the tools of modern airpower. Airpower had not matured to the point of being effective in most situations, especially in its ability to sense ground targets, but the rudimentary basics for resupply and firepower support for potentially surrounded ground troops were becoming a real possibility for the first time in the history of warfare.

The technological basis for airpower to evolve and mature to such a capability and to be applicable to such a concept as the operationally offensive, tactically defensive operation clearly provided new possibilities and increased military capabilities. While the idea of strategic airpower and Guilio Douhet's proposals indicate that some gave thought to what the strategic role of airpower would be, there was no clear linkage between airpower as a tactical advantage and operational facilitator and its strategic impact in a counterinsurgency role.[85] In truth, such a relationship was probably not possible by the 1950s. Militaries were still struggling with the concept of revolutionary warfare and how to deal with it in an effective counterinsurgency manner. Until counterinsurgency operations had a solid foundation of doctrine and theory, implementing airpower in a harmonious way at the tactical, operational, and strategic levels would be impossible. Sadly rather than address the strategic issue of firepower within the role of counterinsurgency, it was far easier to focus on the role of airpower within a conventional setting.

In the 1950s the historical narrative clearly indicated that a conventional conflict between the United States and the Soviet Union—that is, the

North Atlantic Treaty Organization (NATO) and the Warsaw Pact—was a far greater threat and likelihood in geopolitical terms than were small wars with counterinsurgency aspects. It was completely reasonable that thinkers at the time focused their energies on understanding and improving conventional capabilities given the historical context, but they also should have given some consideration about counterinsurgency as revolutionary warfare had such a potential to disrupt governments around the world. The failure to adequately understand counterinsurgency warfare and the role of airpower in it would neither be understood nor explored at that time.

By the 1960s airpower had indeed advanced by leaps and bounds as new technical capabilities were integrated in the U.S. Air Force and new platforms developed for the delivery of nuclear and conventional weapons. As the technical tools of the U.S. Air Force increased, so did its killing power and its ability to deliver ordnance accurately in support of ground forces.[86] This progression would have the effect of making the operationally offensive, tactically defensive operation a greater possibility. Larger amounts of firepower could be delivered more accurately in support of ground troops, and resupply could be carried out under the assumption that air-delivered ordnance could provide the standoff required for the aircraft's use of runways. While technology could not dictate the weather, with increased standoff around the airfield of a defending force the ability to resupply by parachute became a much more reasonable proposition. The strategic realm, however, would still continue to elude the U.S. military and its effective application of an operationally offensive, tactically defensive concept.

As airpower came into its own, especially as a separate service within the U.S. military, it attempted to define itself as an equal to the other branches of service. In part this separation is a natural result of any bureaucratic entity trying to be taken seriously and looking after the organization's own best interest. Champions of airpower, such as Douhet, sought to define airpower as a strategic instrument capable of identifying and quickly reaching the outcomes of conflict through strategic bombing. The example of Dien Bien Phu would be likely to reinforce the airpower enthusiasts' view that airpower is the key to tactical success in an operationally offensive, tactically defensive operation. In part they would be right so far as troops on the ground, to compensate for their small number, rely on ordnance deliveries to provide killing power.[87] In truth where the killing power comes from, be it artillery or air-delivered ordnance, is not as important as its accurate delivery. As the technology demonstrated in 1953, and indeed does today and likely will tomorrow, air-delivered ordnance provides support for the ground troops so effectively because it does not require logistics forward of its actual takeoff point, which could be far behind friendly lines.

Airpower is essential for the effective implementation of an operationally offensive, tactically defensive concept. There is no substitute for airpower with its extreme flexibility and low logistical cost forward; however, it is necessary to remember that the operationally offensive, tactically defensive concept only works with boots on the ground. Forces must be on the ground and capable of staying in the area in order to induce an enemy to attack and spend precious resources in trying to dig the defensive forces out of their position, thereby eliminating the threat posed to their rear area, lines of communications, or center of gravity.[88] So while airpower is an indispensable feature both for the mobility of troops and for supplies, as well as for the delivery of crucial supporting fire, it only has a role within the operationally offensive, tactically defensive operation as support for those troops actually engaged on the ground. Airpower, in this context, should be standing by to provide all support possible, including delivering ordnance, collecting intelligence through sensors, resupplying and evacuating troops, and when possible, effectively interdicting the lines of communication of a besieging enemy force.

VI. Khe Sanh

The French originally constructed the Khe Sanh combat outpost during the colonization of Indochina, and U.S. Army Special Forces occupied it in 1962.[89] By 1966 the outpost had been turned into a sizable base for U.S. military personnel. The base was strategically located in the northwestern corner of what was South Vietnam and in close proximity to the logistics lines used by the North Vietnamese that were commonly referred to as the Ho Chi Minh Trail. Much like Dien Bien Phu, Khe Sanh occupied an area surrounded by high peaks or hills that looked down on the base.[90]

The rationale for occupying and expanding the combat base at Khe Sanh was to provide a base capable of threatening the logistics lifeline to the Vietcong operating in South Vietnam. While technically Khe Sanh was not behind enemy lines and was located within the perimeter of South Vietnam, the entire area around it was decidedly not a friendly area as it was controlled by North Vietnamese soldiers and sympathizers from South Vietnam. The logical course of action for the U.S. military was to exert its influence in the area surrounding Khe Sanh and weaken the Vietcong guerrilla forces in southern Vietnam by interdicting supplies moving south along the Ho Chi Minh Trail.[91] Whether the strategy would prove decisive was not known at the time, but the location offered the best opportunity for an interdiction effort by ground forces.

Khe Sanh represents an excellent example of the operationally offensive, tactically defensive concept in terms of the geographical location and in the context of the enemy's control of the surrounding area. Implementing such a concept in an operation that will be scrutinized nationally engenders tremendous risks to the soldiers involved and political risks to the outcome of the battle.

The situation of Khe Sanh may at first glance have appeared to be almost exactly like that of Dien Bien Phu. Both locations were dependent upon an airfield surrounded by hills and defended by infantry. Dien Bien Phu and Khe Sanh were also similar in that they required resupply and support by air rather than ground due to hostile control of the surrounding areas. Further while requiring airlift for their survival, both areas faced uncertain weather, which hampered resupply and evacuation efforts, and antiaircraft artillery, which posed a threat to the aircraft involved. Khe Sanh had some very important differences from Dien Bien Phu; primarily, the former had airpower to deliver ordnance in close air support and sensing technology to give a general idea regarding the location of enemy forces.[92]

In terms of actual numbers roughly five thousand to sixty-six hundred U.S. troops defended Khe Sanh at any given time against a force of approximately forty thousand Vietnamese in the area with twenty thousand assault troops from the 304th, 324th, and 325th Divisions, with the 304th Division's having taken part in the battle of Dien Bien Phu.[93]

These numbers, however, do not tell the full story. Similar to the battle for Dien Bien Phu, the battle for Khe Sanh would first be fought for the hills surrounding the main base and its airfield. The Vietnamese clearly understood that the battle for the hilltops, in this case the six prominent hills north of the Khe Sanh combat base area, would be the decisive objectives in the operation. If the Vietnamese were capable of taking the hills, they would be able to repeat the events of Dien Bien Phu and pound the actual combat base at Khe Sanh into submission. The initial five thousand U.S. military troops in total had to be divided to defend those hilltops. By contrast the Vietnamese were capable of massing forces to assault hilltops of their choosing. Therefore, the roughly 8:1 manpower advantage that may seem to have favored the Vietnamese forces was, in fact, much greater. For example, the battle for Hill 861A was fought at 0300 on February 5, 1967, with elements from the North Vietnamese 325th Division attacking and being repulsed by a single company—specifically, E Company, Second Battalion, 26th Marines.[94]

Many of the hillsides around Khe Sanh saw battles similar to the one for Hill 861A, in which small numbers of U.S. Marines held off vast numbers of attacking North Vietnamese. The U.S. Marines, like the French at Dien Bien Phu, had some armor at Khe Sanh that was used for internal defense.[95] The obvious difference between the American effort at Khe Sanh and the French effort at Dien Bien Phu, which the French speculated earlier to have lost because they lacked airpower, was indeed the Americans' ability to call on air-delivered ordnance. Further, unlike the French at Dien Bien Phu, the Americans at Khe Sanh were able to use methods of intelligence collection ranging from electronic and remote sensing to foot patrols in order to ascertain the enemy's location and effectively use air strikes before enemy forces attempted to close with the defenders.[96] The U.S. forces used these technical tools properly to provide the maximum amount of firepower and support for the infantry. In this sense Khe Sanh represents a truly revolutionary implementation of technology to influence a tactical and operational situation.

Quite likely no amount of effort by the North Vietnamese forces, or massed manpower, would have made them capable of overwhelming the American forces at the combat base at Khe Sanh and in the surrounding hills. This battle is a fairly significant event in the history of warfare. Infantry, when properly supported with the correct technology to bring devastatingly high volumes or properly targeted smaller volumes of supporting fire, had become immune to infantry assaults by enemy forces. Granted the defending force did still need a robust presence on the ground capable of fighting, but in the case of Khe Sanh, the battle was determined before it ever began by the technical capabilities of the defending force.

The North Vietnamese, having remembered the lessons of Dien Bien Phu, were probably confident of their ability to attack and overrun the hills surrounding Khe Sanh and thereby bring the acquired artillery advantage from those hills to bear on the actual combat base.[97] They understood the value of antiaircraft fire as it was effectively deployed and used in the battle for Khe Sanh. They were also very aware, more so than the Americans were, of the detrimental effect of the weather on aerial resupply, evacuation, and close air support. However, basing preparations for a dynamic and new conflict on what was used in the last war rarely serves a force well. In this case the accuracy of high-altitude bombing by the B-52s in support of ground forces, as well as the use of sensor technology in identifying the massing of enemy formations, added a new dimension for which the North Vietnamese were not prepared.

During the battle for Khe Sanh, B-52s flew 2,548 sorties and delivered 59,542 tons of bombs in support of the U.S. Marine ground forces.[98] The B-52s' ability to fly above any weather conditions and provide reasonably accurate close-in air support to within three-quarters of a mile of the base's perimeter proved quite important.[99] They provided an effective ability either to prevent enemy formations from massing prior to an assault on a U.S. position or to devastate them as they did so. In effect the capability to deliver devastating air ordnance from within a quarter mile of U.S. Marine positions by typical close air support to an almost unlimited range, dictated only by the understandings of intelligence and sensors, extended the zone in which time for space could be traded. This technical advantage destroyed General Giap's hope for a decisive military victory at Khe Sanh.[100]

Just as the minié ball in the American Civil War extended the zone in which time for space could be traded over the musket ball by several hundred yards and as the rifle and machine gun in World War I extended the zone in which time for space could be traded by hundreds to more than a thousand yards, so now airpower had been able to extend the zone in which time for space is traded to devastate attacking formations out to the range of the available sensor technology. The B-52 was obviously not the only aircraft or platform capable of doing so, as almost any airborne close air support would have the same effect; but the B-52's ability to operate in any weather conditions and still deliver ordnance effectively made it stand out.

By 1968 the battle of Khe Sanh demonstrated that airpower had truly come into its own in its ability to provide close air support to American infantrymen who were surrounded and in contact with numerically superior enemy forces. The tactical and operational use of the technical means available was cutting edge for its time; however, the technological superiority of U.S. forces, with all the substantial increases in ability to devastate enemy formations, was still not being applied in a strategic manner to make the technology actually decisive. The United States was, and still is, suffering from a strategic deficit in understanding counterinsurgency operations and opposition to revolutionary warfare.

Khe Sanh stands as a testament to the technical innovations possible and the strategic limitations inherent within the U.S. military, at least as of 1968. Technical tools such as airpower provided major tactical but not strategic advantage.[101] In a repeat of French mistakes in Indochina in the 1950s, the United States became involved in a conflict in which it could have prevailed, but it insisted on making the same strategic errors as the French before them by focusing on attrition rather than on politics.[102] While Khe Sanh stands as arguably the finest moment in tactical and operational proficiency of the Vietnam War, it highlights just how far removed U.S. forces were from the minds of the populace and the true center of gravity in a counterinsurgency effort when confronting Mao's revolutionary warfare.

By the end of the battle of Khe Sanh, an estimated two hundred U.S. Marines and some twelve thousand North Vietnamese lost their lives.[103] A 60:1 exchange ratio is a truly remarkable mathematical feat to accomplish in warfare. Such exchange rates are almost unheard of historically, and it should validate the technical prowess and its importance in modern warfare. But focusing too strongly on such a ratio creates the dangers of accepting attrition-style warfare, which is void of strategic thought and innovation.

By 1968 the operationally offensive, tactically defensive concept had become a reality for conventional warfare as never before. The concept's utility is predicated on the assumption of maintaining air superiority for ordnance delivery and for flying close air support and mobility-supply roles. Forces experienced supply shortages at Khe Sanh but not to the extent that the garrison was ever threatened because they maintained air superiority and held the surrounding hilltops.[104] Conventionally no strong evidence indicates that the concept could not be applied at any point and at any time against the technically inferior foe, who was unable to contend for air superiority.

It is important to note that sensor technology in 1968 was in its infancy, as were precision-guided munitions.[105] Long before the current debate on the revolution in military affairs arose, there was a revolution in military organization relating to the combat effectiveness of infantry units working jointly. Of course this effectiveness is defined within a conventional framework, but the operationally offensive, tactically defensive concept, and the technologies supporting it, do not have to remain within the realm of conventional warfare.

In the years after the battle of Khe Sanh, the United States would have to come to terms with the reality of winning operational and tactical engagements, almost without exception, in Vietnam while utterly losing, in spectacular fashion, the political battle being waged. The logical conclusion would dictate taking the lessons learned and molding and disseminating them throughout the ranks so that they could better understand the mistakes made at tactical, operational, and strategic levels and not repeat them. This process would require an understanding of counterinsurgency operations with the expectation of having to implement those operations in future communist-inspired revolutionary wars. To some extent this retention of knowledge did happen, specifically in the special operations community in the U.S. military, but in the conventional side of the services it is not the norm.[106] The technologies that were so effective at the battle of Khe Sanh continued to evolve, with a focus on the conventional realm of warfare. However, neither the technologies nor the strategic thinking would be integrated in the U.S. military's plans regarding how to effectively engage in counterinsurgency operations, and the impact of the resulting strategic deficit would last thirty years.

VII. The Infantry

The operationally offensive, tactically defensive concept revolves around the participation and support of the infantry in a defensive position.[107] Obviously the infantry is capable of patrolling and being active in defending the area, but generally speaking, the area will probably be in the hands of enemy forces or influenced by enemy forces so the overall posture of the infantry will be defensive in nature.

In the American Civil War example, Longstreet's concept was based on the infantry's effectiveness in defensive positions at the battle of Fredericksburg. Trench warfare in World War I clearly was predicated upon the ground held in trench systems used by the infantry, demonstrating the extreme difficulty of digging out well-fortified units. The battles at Dien Bien Phu and Khe Sanh, while having drastically different outcomes, were both centered around defensively entrenched infantry forces holding territory and the use of supporting fire to protect them. The operationally offensive, tactically defensive concept can only work with an infantry component as the core of the defensive structure. Whether this infantry structure will be mechanized infantry or light infantry will depend on the mission's objective and the terrain to be held.

The improvements in the technical capabilities of infantrymen have been substantial since the 1970s. Other than the fact that a soldier still fires a 5.56mm round from a shoulder-fired individual weapon, today the U.S. Army and Marines issue their soldiers gear that is almost wholly different from that used during that time.[108] The infantry itself as an organization and the tools that it uses are evolving just as quickly as the technical assets that are available to support it. Arguably this evolutionary, and perhaps sometimes revolutionary, progress in technical capabilities will continue into the foreseeable future, making both the infantry and its supporting elements more capable of fulfilling their missions but also logistically overburdened.[109]

The examples listed of the historical precedents of the operationally offensive, tactically defensive concept of warfare show a clear trend in the numbers of infantry involved. When General Longstreet proposed an operationally offensive, tactically defensive concept to General Lee during the American Civil War, he likely envisioned the entire Army of Northern Virginia as the element that would entrench itself into that defensive position. This entrenched force of roughly seventy-five thousand Confederate soldiers would have required an enormous attacking force to overwhelm and annihilate.[110] By 1953 the French Union forces, if they had had adequate airpower in a supporting role, could have held the base at Dien Bien Phu with an estimated maximum strength of fifteen thousand soldiers.[111] By 1968 in the battle of Khe Sanh, five thousand to sixty-six hundred American soldiers were capable of holding a defended fortified area, with the availability of supporting fire, with a 60:1 casualty ratio.[112] The trend here should be clear: as the application of technical means of delivering firepower increases both in quantity and in quality, the subsequent requirement for the number of soldiers defending a given objective decreases. In short, technology becomes a direct force multiplier for infantry when involved in operationally offensive, tactically defensive operations.

The stated trend can obviously be questioned on the grounds of the number of troops that each of the defending forces faced. For example, the Army of the Potomac under Gen. George Meade, although not vastly superior in number, had perhaps twice as many combat troops as those troops available to the North Vietnamese assault forces in the battle of Dien Ben Phu. If the North Vietnamese had had the same number of troops that Meade had available, then arguably the North Vietnamese would have had a larger advantage over French Union forces. But the same would not be true if the number of troops available to the North Vietnamese for the battle of Khe Sanh had been similarly increased because, as noted before, the number of troops was no longer the determining factor of whether an offensive force could overrun the position of a defensive force. By the battle of Khe Sanh, the number of troops involved became an afterthought to the amount of firepower and an effective targeting of the firepower in support of the defensive units. So while each of these historical incidents took place at different times, in different places, against different forces, and with a different number of opposing forces with different technical capabilities, the trend from the past should hold true into the future: the number of infantry involved in operationally offensive, tactically defensive operations will decline as the technical capability available increases, assuming technical asymmetry.

Of paramount importance to this book, as it is concerned with counterinsurgency doctrine, is a linkage of this operationally offensive, tactically defensive conventional concept to an application in the field of counterinsurgency. In the historical examples, technology allowed the infantry to operate in smaller and smaller units. The combination of increased fire support and the technical means of intelligence to locate enemy targets had increased to the point of making company-size units on the hills surrounding Khe Sanh impervious to infantry assaults. It should be noted that it took a substantial amount of airpower to make that infantry's line impenetrable, but nevertheless, the technical means were achieved. However, by 1968 the technology required to move the operationally offensive tactically, defensive concept into a practical application for the counterinsurgency realm was not yet mature enough to be effective.

Following sound counterinsurgency doctrine, for the infantry to effectively conduct counterinsurgency operations through the application of the operationally offensive, tactically defensive concept would require dispersing infantry units, each consisting of few soldiers, to outlying villages to secure them from insurgent forces.[113] However, while support and especially air-delivered ordnance for close air support had matured, the survivability of the infantryman on the ground had not yet reached a point where extremely small units could withstand an attack by enemy forces that had managed to move close enough to engage in direct small arms fire. This fact alone likely would have prevented the dispersal of small infantry units because of the near certainty of them being defeated in detail by larger North Vietnamese formations.[114] It should also be noted that as the infantry is broken down into smaller and smaller units to use the operationally offensive, tactically defensive concept at multiple points, the potential requirements for greater amounts of close air support, as well as the requirements for timely intelligence for each of the small units, will increase exponentially.

The dispersal of infantry units into small, reasonably independent forces should be a prerequisite for the effective implementation of the operationally offensive, tactically defensive concept in counterinsurgency operations. By recognizing that the center of gravity of the conflict is in the minds of the populace rather than in the military structures of the insurgent, the infantry's ideal tactic, operation, and strategy should be focused on securing the population in the promotion of good governance at all levels while using highly specialized forces to track and kinetically engage insurgent cells, leaders, and formations.[115] This winning over the population can only be done by forces on the ground in each village, town, or small city. Because the minds of the people are paramount in the struggle and because they must be secured, forces would be required in every location that has a population center. Historically, without vast numbers of troops and a willingness to take a vast number of casualties and to accept the occasional defeat in detail of friendly units, the infantry has not been able to provide this role. For example, if it required five thousand to six thousand U.S. troops to make one location secure and impervious to enemy attack, as in Khe Sanh, it would theoretically require the same number of U.S. troops in every Vietnamese village to guarantee total security for the Vietnamese population in 1968. This notion is ridiculous and clearly beyond the capacity of any military to do, and it ignores the basic principle of allocating forces based on intelligence; but mathematically it does illustrate the problem of providing theaterwide population-centric security.

The shift required for the infantry to effectively carry out counterinsurgency operations in the application of the operationally offensive, tactically defensive concept requires a change in the military's mind-set to allow the dispersal and independent action of infantry units. Organizationally the very nature of the U.S. military culture will strongly resist such shifts.[116] As stated the technology has matured, especially the technical capabilities for providing support and intelligence, as well as for enhancing the survivability of infantry units. The capability of the infantry to operate in smaller units has become a reality.[117] However, the typical mind-set in the conventional military has not changed significantly. This is not to say that the U.S. military is incapable of adjusting or adapting to new tactics and strategies to meet the demanding and dynamic changes that may confront them. As the U.S. military proved in Iraq in 2007, it is capable of adapting at the strategic level given the appropriate leadership.[118] The U.S. military now engages in counterinsurgency warfare using infantry as the focus for securing the population in pursuit of the strategic and operational goals. Yet, as of 2008 in Diyala Province, Iraq, the U.S. military was not comfortable dispersing its units to villages to secure as much of the population as possible.[119] Rather, it preferred a more slow-moving conventional approach of accepting insurgent control over vast swaths of territory and attempting incrementally to bring areas under the control of counterinsurgent forces.

Iraq has shown that the U.S. military has finally started to address the strategic deficit in counterinsurgency operations through the application of quasi-sound doctrine. Allowing the infantry to secure the population and to interact with it in gathering intelligence is a step forward and in the right direction. While it should not have taken thirty years since the failed counterinsurgency operations in Vietnam to adopt these measures, any step forward is a positive one; but the process of evolving the U.S. military into a confident counterinsurgency organization capable of using all the technical tools at its disposal to be as effective as possible on the tactical, operational, and strategic planes has not yet been achieved. The technical tools are now present, the center of gravity identified, and the operationally offensive, tactically defensive doctrine mature for the U.S. military to synergize all three aspects and implement effective counterinsurgency operations based on the infantry. The only question that remains is whether the U.S. military is willing to take the organizational, political, and financial steps required to implement a new way of carrying out counterinsurgency operations, one that focuses on securing the population en masse in order to strike decisively at the center of gravity of the conflict and yield strategic results rather than taking incremental steps in given areas that allow the insurgents to retain effective control of most areas of operation.

VIII. The Future

If the U.S. military can take proper advantage of the technical capabilities of existing force structure and the potential of technologies emerging from research and development to facilitate the operationally offensive, tactically defensive concept at the lowest possible infantry level, then the outcome will be profound. The future of such an endeavor would define theaters of operation in which counterinsurgency warfare was the norm and not nearly as difficult a problem as it has been perceived historically.[120] Such a new method of counterinsurgency warfare will provide tools at the strategic level for policy makers to feel more comfortable about using hard power and thereby the likelihood of implementing a counterinsurgency approach to failed or threatened states. Any state or non-state actor within the state system would have to take such a military capability by the United States very seriously. For too long it has been an accepted truth that conventionally minded, technically driven Western powers cannot effectively combat locally driven and indigenous insurgent movements. Only a state that would be completely sure that its population would stay loyal, or a non-state actor that was absolutely certain that its message was so universal that the populace was guaranteed to side with it, could afford to discount U.S. military power.

The logical and predictable path for an insurgent group that did not carry out an insurgency to win popular support, or threaten the government's ability to provide security in any given location, would be to resort to terrorism in an effort to convey its political dissatisfaction and demands. This book does not develop or delve into the role of terrorism other than to accept that insurgents commonly use it is a tactic. As it does not win the support of most populations, terrorism likely is to be neither of long-term benefit to the insurgent nor a deciding factor at the tactical, operational, or strategic level for counterinsurgency forces; rather, it only serves to garner attention for a specific cause.[121] Just as important, the employment of terrorist activities and methods by insurgent forces would not detract from the extremely powerful position the United States would attain by exploiting its ability to engage in counterinsurgency warfare and the widely accepted perception that it is capable of doing so with an extremely high likelihood of success.

If the U.S. military were to succeed in changing the fundamental perception that counterinsurgency operations can be effective, that in itself would be a tremendous victory. Owing to the failure of many attempts over the past sixty years, the prevalent pessimism over the ability of Western powers to effectively engage in counterinsurgency operations needs to change if the U.S. military is to continue to have the same level of deterrent and potential power as it has had in the conventional realm over the last century. Successful counterinsurgency efforts in Iraq until the U.S. forces' withdrawal, and possibly Afghanistan, depending on the results of both operations, may go a long way to dispelling this myth.[122] If successful, the United States will have achieved it by applying *basic* strategy and using technology and technical means to help implement that strategy. However, it is unlikely that the present use of the U.S. military can accomplish it. Nevertheless, the United States has the potential to redefine its military capabilities by focusing technology on counterinsurgency success. Such redefinition will have resounding political implications and provide future policy makers with the strategic flexibility to use force when needed.

Becoming complacent and accepting a single mode of operation or a single set of technical capabilities in the pursuit of counterinsurgency objectives are not in the interests of the United States. Insurgent forces are constantly seeking new ways to improve their performance and achieve their objectives.[123] It is imperative that the U.S. military continues to do the same through examining doctrine, applying new technical means and capabilities, and maintaining its flexibility in operations.[124]

The operationally offensive, tactically defensive concept has a future in counterinsurgency warfare, providing a continued investment in the evolution of technical capabilities of the U.S. military. It represents the best-option for the integration of technical means to enhance military capability while maintaining harmony and synergy between the tactical, operational, and strategic levels. Research and development will enhance the technical means for soldier survivability, mobility, close air support, supporting fire, and the technical means of intelligence gathering and analysis, which, in turn, can be provided down to an extremely low level— that is, to the squad or platoon. In doing so the operationally offensive, tactically defensive concept would be well suited for counterinsurgency warfare. As forces operationally move into large swaths of enemy-held terrain and become embedded in small villages or towns, they represent a direct threat to the center of gravity of the enemy. This threat is grave enough to the insurgents' desired political ends that just as General Longstreet was able to predict the North would not tolerate the Confederate Army in Northern territory during the American Civil War, the United States can presume that the insurgents will not want small units of American soldiers providing security over large swaths of territory in what amounts to the insurgents' lifeline, or their single most important center of gravity.[125]

In such an application of the operationally offensive, tactically defensive concept, it can be expected that insurgent forces will exercise the maximum amount of pressure in as many different locations possible. The insurgents will have to overrun and defeat individual outposts of U.S. military personnel in order to relieve the strategic pressure on the center of gravity and force the United States to reconsider the operationally offensive, tactically defensive concept. If the insurgents are capable of inflicting enough casualties—and casualties will occur and could be high in the beginning—political and financial costs could be steep enough to threaten the viability of the concept at the political level. Any large-scale coordinated attack designed to overrun and defeat small units, however, could also be seen as a validation of the operationally offensive, tactically defensive concept. Clearly while insurgents will attack any target they see as a viable option, generally speaking they will attack those targets that provide an almost certain guarantee of success.[126] Consider the words of General Giap, who commanded North Vietnamese forces against both the French and the United States, when recalling the battle of Dien Bien Phu: "We came to the conclusion that we could not secure success if we struck swiftly. In consequence, we resolutely chose the other tactics: To strike surely and advance surely. In taking this correct decision, we strictly followed this fundamental principle of the conduct of revolutionary war: Strike to win, strike only when success is certain; if it is not, then don't strike."[127]

For an enemy force to strike dispersed infantry U.S. military units, which are specifically postured to defend their position against attack and have access to commonly available joint assets such as overwhelming and devastating close air support, would still indicate the insurgents' desperation and their concern for the center of gravity. It is quite unlikely, unless U.S. forces were continuously overrun, that insurgent attacks on U.S. forces would fit General Giap's formula for success.

The real technical test of the operationally offensive, tactically defensive concept as applied to counterinsurgency warfare will come down to the effectiveness of the technical sensor means in intelligence collection that is available to the dispersed units. As noted previously, the operationally offensive, tactically defensive concept relies on the ability of the infantry to trade time for space out to the maximum distance possible against an encroaching opponent. It is wholly insufficient to accept that the distance of time and space that could be traded is limited to the kinetic projectile fired by the infantry. It is absolutely imperative that the infantry unit has a specific integrated capability to understand and sense the terrain around it.[128] A realistic range of ten miles of sensing radius around an infantry unit, with the attendant technical means to deliver accurate precision-guided munitions to destroy those enemy targets, would allow a large enough window for trading time and space so that the actual size of the infantry unit could become very small.[129] Ideally ten soldiers would be able to effectively hold terrain and secure the populace against overt aggression by insurgent formations. Every mile that an insurgent would move toward the defending infantry also represents a mile he would have to move, likely by foot, away from the defending infantry. As the insurgent moves closer and moves away, he becomes vulnerable to attack from ever-increasingly accurate and responsive fire called or controlled by the infantry.

While the operationally offensive, tactically defensive concept for implementing counterinsurgency warfare would rest on technical capabilities to continue the trend of minimizing those infantry forces capable of defending an area, there would also be strong pressures for the soldiers to master skill sets that would facilitate the concept. Many of these skill sets can also be enhanced through technical means. For example, for an infantry unit in a village surrounded by hostile insurgents, technical intelligence alone will not suffice. The soldiers will have to foster human intelligence (HUMINT), which could play a decisive role as well. (The specific technical means to accomplish this goal is addressed in the following chapters.) Here, however, the responsibilities of a ten-man unit would appear to be more along those of a traditional U.S. Army Special

Forces Operational Detachment Alpha (ODA) rather than a pair of conventional infantry squads.

The challenges associated with implementing the operationally offensive, tactically defensive concept would be manifold but not beyond the capabilities of the U.S. military either to field today or to reasonably develop with research and development for the near future.[130] Most of the connectivity, bandwidth, sensing technologies, and other technical means by which trading time for space could be achieved for fielding smaller units in a dispersed fashion have been developed. What is lacking is a coherent and secure network specifically designed to support their use by troops employing the operationally offensive, tactically defensive concept in a counterinsurgency role. This shortcoming, of course, means that finding the solution is an engineering problem, not a conceptual problem, and given enough time and financial resources, almost all engineering problems are surmountable.

As the United States attempts to engage in counterinsurgency warfare, it must understand that it will have to be flexible and dynamic in response to asymmetrical challenges made by insurgents. Reflecting a basic premise in warfare outlined by both Sun Tzu and Clausewitz, such flexibility and adaptability will require audacity if the United States is going to be successful.[131] Counterinsurgency, by its very nature, has to deal with asymmetric challenges and be innovative, not complacent. It cannot fall back on what seems comfortable as the insurgent will surely not cater to those wishes. The operationally offensive, tactically defensive concept provides a framework that allows the United States to use its technical expertise and excellence in a proper capacity and further its strategic aims. The concept permits flexibility for policy makers when they examine what lies within the military capability of the United States. The concept further drives the operational and tactical initiatives to fall into line with sound counterinsurgency practice and theory. Finally, the historical examples of the use of the operationally offensive, tactically defensive concept show a continuous trend of integrating technical tools, thus allowing for the reduction in the number of forces needed. Obviously the technical disparity between both sides will give an increased advantage to the defender; it is a crucial requirement that the defender should possess technical superiority or parity at a minimum. Technical superiority directly plays upon the strengths of the U.S. military and represents the military's greatest opportunity to leverage its technological base and produce strategic effects in counterinsurgency operations by adhering to sound counterinsurgency practice.

2   Counterinsurgency in Iraq

> Time is like a river made up of the events which happen, and a violent
> stream; for as soon as a thing has been seen, it is carried away, and
> another comes in its place, and this will be carried away too.
> MARCUS AURELIUS

I. 2003 Invasion

The 2003 U.S. military intervention in Iraq fundamentally challenged the
U.S. military to adapt and obtain U.S. policy objectives in a counterinsurgency conflict, which was far removed from the conventional context
in which it was prepared to fight during the Cold War. While unconventional warfare is not new to the United States, with its force structure
and doctrine in 2003, the military was neither prepared for nor expecting to find the level of tenacity and competence of the fundamentalist
Islamic forces and Fedayeen Saddam in Iraq that became apparent after
the fall of Saddam Hussein. Such a failure in perspective on the part of
U.S. forces regarding counterinsurgency is somewhat hard to understand
given historical predictions that U.S. military personnel would face that
type of conflict.

The technical capability of U.S. military hardware allows it an overwhelming advantage on the battlefield. Indeed, U.S. forces have actively
fostered the technology used on the battlefields today. It is somewhat
confusing and difficult to explain then how a military force that has
put such emphasis on the technical capabilities of its own military force
structure could expect that future adversaries would either be ignorant
of such advantages or play into them. Throughout the 1990s there was
considerable speculation that future conflicts would be counterinsurgen-
cies.[1] For the U.S. military to enter into a conflict in Iraq in 2003 and be
surprised at facing an insurgency rather than a sustained conventional
force is negligent at best.[2]

There was little question concerning the U.S. military's capability to defeat the armies of Saddam Hussein in Iraq in 2003.[3] Iraq's conventional forces were insufficiently capable of withstanding a conventional assault by U.S. forces; however, the United States then found itself poorly prepared and equipped and surprised by a homegrown Iraqi insurgency. The pertinent issue is why the U.S. military was convinced that a conventional victory on the battlefield in 2003 would end hostilities and lead to the formation of a new government by the Iraqi people and to swift reconstruction. The relevance of this question directly impacts on the viability of U.S. military intervention in the future and highlights both the capabilities and the inadequacies of U.S. military technology.

This chapter is not concerned with the actors and infighting between policy makers who were responsible for the development of strategy for Iraq. Their competing visions and advocated positions are not relevant here. This discussion is only concerned with the consequences of the actions taken from 2003 onward with the attendant challenges for the U.S. military to conduct counterinsurgency warfare.

The U.S. military should have anticipated a counterinsurgency struggle after the almost certain success of the opening conventional phase of the 2003 invasion. Even if the Iraqi people, as a whole, generally supported the removal of Saddam Hussein from power, obviously a power shift would have had to take place in Iraq in order to have a democratically elected government. A cursory glance at the demographic makeup of Iraq, with a majority 60–65 percent Shia population ruled by a minority Sunni government, should have indicated that the seeds of conflict were present.[4] It should have seemed inconceivable that in a tribal Arab society the Sunni minority would ever peacefully agree to the redistribution of its power to the Shia population through a democratic process. The end result was a lack of planning and the formation of an inept strategy that was based solely on hope and enthusiasm, neither of which represents a technique.

The formation of strategy and the implementation of that strategy using military force are not an exact science. It is impossible to break strategy down into ones and zeros to be crunched and calculated in a mathematical formula that provides predictable and repeatable outcomes. The desire of policy makers and military officers to have predictability concerning military events can be extremely detrimental. When a rigid force structure faces a dynamic insurgency, the former should expect unpredictable variables, unforeseen outcomes, and adaptation by the insurgents. All levels of government must recognize that an effective strategy must be developed, with the ends, ways, and means to achieve it, *prior* to the intervention by U.S. military forces. Lt. Gen. Michael DeLong, the deputy commander in chief of Central Command, insisted that it was his idea to hang a Mission Accomplished sign behind President George W. Bush as he gave his Iraq War speech in May 2003 to signal the end of major combat operations and to persuade European countries to enter the Iraq conflict.[5] However, the president of the United States should have had the foresight to see the dangers of such a political move to announcing the end of hostilities. The desire for well-defined solutions to counterinsurgency conflict, and the easily defined ends, exemplifies a general tendency by Americans for order in an endeavor that is inherently chaotic.[6]

As previously stated, generalizations about American culture are dangerous and lead to a slippery slope. Given the overwhelming number of different cultures and cultural backgrounds of the people residing in the United States, any effective generalization will have exceptions, sometimes glaring ones. Culturally speaking, ambiguity in warfare is a hard concept for Americans to accept. Such ambiguity, rather than rigidity in morally and culturally driven doctrine, is part and parcel of the very essence of counterinsurgency practice.[7] Any strategy to confront an insurgency and effectively counter it in support of a host nation will require a flexibility that is at odds with the American cultural norm.

Some American people were supportive of the invasion, even though they did not have a clear understanding of the possible consequences regarding the development of an insurgency. Many Americans followed the invasion live on CNN as embedded reporters beamed back reports of the conventional forces' advance on Baghdad. The enthusiasm was short lived as scenes of wholesale looting in the capital soon followed. It should have been clear from the looting that Americans lacked a fundamental understanding of Iraqi society. Many Americans voiced bewilderment at the absence of civic virtue as crowds overran and looted government agencies, which were going to be needed if Iraq was going to have a functioning representative government. The American people at many levels—congressional leaders, the Bush administration, and the U.S. military—held simply inaccurate assumptions.[8]

From a certain perspective, the U.S. military cannot be held accountable for the failures of its commander in chief; however, it can be held accountable for its inability or unwillingness to give sound advice to the commander in chief regarding the best possible deployment of U.S. military power. It is absolutely natural that planners could have made certain assumptions as the previous conflict in Iraq in 1991 to liberate the state of Kuwait, the Gulf War, was between conventional forces using conventional doctrine.[9] Further as the U.S. military often does, it trains for and expects to fight the last war. The intervention in Iraq in 2003, however, was clearly going to be different from the intervention to free the state of Kuwait in 1991. Removing the Iraqi Army from Kuwait did not have to take into account a possibly hostile indigenous population. Therefore, in 1991 no real likelihood of an insurgency needed to be planned for in order to successfully accomplish the political objective of expelling Saddam's armies from Kuwait. While an insurgency might have been on the minds of some thinkers at the time, it would have been led by anti-Saddam Shia insurgents in the marshes of southern Iraq. Therefore, any potential insurgency might have been supportive of the U.S. objectives of limiting the power or weakening the armies of Saddam Hussein.[10] There was a reasonable expectation that a force-on-force confrontation in 2003, as U.S. forces moved north of Kuwait and into southern Iraq, would be similar to that between Iraq and the coalition forces in 1991. If anything, planners expected the 2003 coalition forces to win a more overwhelming victory.[11] By 2003 technology, improved force structure, and doctrine emphasized mobility and information, providing the same combat effectiveness by fewer troops than in 1991. The overwhelming military technical advantage of the United States and its coalition allies in 2003 demonstrated the advantages flowing from technical superiority.

Predictions of Iraqis greeting U.S. and coalition forces as liberators were the best scenario or outcome for the United States. Hoping for the best should never be central in planning the conduct of a military intervention; the old axiom that one plans for the worst and hopes for the best is a far better tool than simply planning for the best. While undoubtedly large segments of the population in Iraq saw the removal of Saddam Hussein from power as a great advantage and a worthy goal, the later emergence of an insurgency, particularly among the Sunni Arabs, should have been expected.[12] The fall of Saddam Hussein and the Ba'ath Party was going to have ramifications for the social fabric and distribution of power in Iraq.[13] To achieve military and political objectives, such issues should have been on the minds of military thinkers if they had been thinking strategically and not only operationally or tactically. This lack of forethought on the part of the executive branch of the U.S. government created false expectations and hopes that a military intervention in Iraq in 2003 could be quickly and decisively accomplished at a minimal cost in lives, treasure, and political capital.[14] While it is very easy to criticize the decisions that were made at the time, especially with the advantage of hindsight, clearly, the situation was not well thought out prior to the intervention.

Counterinsurgency has not been central in U.S. military thinking since the Civil War. The defense industry is centered around conventional requirements, the bureaucratic machine of the Department of Defense is conventionally minded, and even think tanks have been generally gravitating toward the revolution in military affairs and more conventional issues. Most of the thinking coming out of the Office of Net Assessment since the mid-1980s has focused on conventional forces and the possibility of a conventional confrontation between the U.S. military and the conventional forces of a foreign country.[15] During the Cold War, this confrontation would have focused on Soviet and U.S. forces in Europe. With the 1991 intervention by U.S. and coalition forces, under the auspices of the United Nations (UN), to liberate the sovereign country of Kuwait from Iraqi military forces, the focus on conventional warfare seemed to have been validated. In the same decade as the intervention in Iraq and Kuwait in 1991, U.S. forces waged other engagements, most notably in Somalia, that were not conventional.[16] These actions should have been widely recognized as an indication that counterinsurgency warfare would be an issue for U.S. forces in the decades to come. Granted it is easier to theorize about the nature of future warfare than it is to actually prepare, equip, train, and field an army so that it is capable of handling a counterinsurgency conflict. The evidence is clear that in the 2003 intervention, the U.S. Army then believed that it would have a similar experience to the 1991 Gulf War.[17] The true ramifications of the 2003 intervention in Iraq were not foreseen by the U.S. executive branch prior to the insurgency. Otherwise, the prestige of the United States and the Office of the President of the United States would not have been so badly tarnished by President Bush's premature assertion that the mission had been accomplished.

There are notable conventional differences between the 1991 intervention to liberate the state of Kuwait and the 2003 intervention to remove Saddam Hussein's regime from power. While the 1990s saw much interest in the revolution in military affairs, its actual implementation by the U.S. armed forces was a very slow and drawn-out process. The reorganization and metamorphosis of a large force structure cannot happen overnight; in fact, one should expect it to take many years. However, there were efforts to modernize the U.S. armed forces toward the "objective force," which is supposed to be smaller, more agile, and logistically easier to maintain in the field.[18] The forces used in 1991, when compared with the forces used in 2003, clearly show this trend. While the United States and its coalition allies fielded more than 500,000 soldiers and massive amounts of equipment to liberate Kuwait in 1991, they fielded fewer than 170,000 soldiers to invade Iraq in 2003 and remove Saddam Hussein from power.[19] In 2003 the reduction was a very controversial decision by Secretary of Defense Donald Rumsfeld, who felt that a smaller, lighter, and faster force would be more effective and bring more firepower to bear than would a larger force.[20] But this option raises the question whether the executive branch understood that its forces were going to have to occupy the country and rebuild a functioning government in Iraq as opposed to just defeating the Iraqi military. Yes, it is true that a smaller, lighter force, especially with new technologies integrated into that force structure, was capable of defeating the armies of Saddam Hussein much more rapidly than did the large, cumbersome force of 500,000 soldiers in 1991. It is equally clear, however, that such a small and agile force was not going to be adequate to provide security throughout the Iraqi state as would be required by sound counterinsurgency strategy.

Clearly expectations were that the Iraqi government would continue to function after U.S. forces invaded and removed the Ba'ath Party from power.[21] Planners presumed that the police would continue to function in the role of policing civil society and ensuring security at a local level. They also assumed that the Iraqi military would continue to function and provide security at a broader level; however, its dissolution by Paul Bremer, head of the Coalition Provisional Authority, left a security gap. Scenes of looting after the fall of Saddam Hussein's government in Baghdad were beamed all over the world. There was no doubt that the Iraqi government had ceased to function and that the security services were unable to maintain an adequate level of security in Baghdad. Rampant looting went unchecked by Iraqi security forces, which were unwilling or unable to police, and by U.S. military forces, who were unwilling or unprepared to take over the policing duties of the Iraqi security forces.[22] The end result was an evaporation of security both in the Baghdad area and throughout Iraq in general. In turn it led to anger and resentment among the Iraqis. The chaotic situation was worse than it needed to be and proved an ideal growth medium for the Iraqi insurgencies.

In the U.S. military's defense, with its 170,000 soldiers in country, it was unable to provide security in every city and every province throughout Iraq. Coalition partners also were not present in large enough numbers to effectively provide the level of security required to bring the situation in Iraq under control. The highly capable firepower-centric and very mobile technical force that the United States fielded, though more than capable of defeating a conventional force in the field, was ill suited for keeping the peace in a civil environment. There should have been no reasonable expectation that the United States and its coalition allies could police an entire country's civil population and borders with the troops available in Iraq in 2003. Understandably planners had relied on the notion that Iraqi security forces would continue to operate and stabilize the situation in the country until a new civil government was formed.

II. Technology in Iraq

The U.S. military of 2003 was drastically different from the U.S. military of 1991. It was fielded with many of the same technical capabilities, but they had been upgraded substantially. Meanwhile the years of sanctions since 1991 had drastically reduced the combat effectiveness of the Iraqi military.[23]

There was never any question about the outcome of a conventional conflict between the Iraqi military and the U.S. military and its coalition allies. In 1991 the use of precision-guided munitions was well publicized by the Pentagon and often aired during the conflict as a demonstration of the U.S. military's technical capability.[24] By 2003 the use of precision-guided munitions had become commonplace. The weapons systems available to the United States, in fact, had grown more sophisticated and more varied in the intervening twelve years. The dramatic increase in the effectiveness of body armor for U.S. troops, as well as for vehicles in the U.S. force structure, had evolved into protection better than what had been available in 1991. The ability to process data and transfer that data out to actual commanders in the field was now routine throughout the military, and the use of blue force tracking technology in military vehicles easily allowed the tracking of U.S. combat forces, increasing situational awareness throughout the force.

In 1991 the primary use of precision-guided munitions was through laser designation for air-delivered ordnance.[25] This method of delivering ordnance is extremely effective in clear weather but can be severely limited by atmospheric conditions, such as clouds and dust storms. In short, laser-guided munitions are not capable of providing close air support or strategic and operational employment in all weather. The use of laser designation can also put U.S. forces at risk, especially if that designation has to be done on the ground by special operations teams operating behind enemy lines. U.S. Special Operations Forces in the western desert of Iraq in 1991 had to use this method in an attempt to limit the number of Scud missiles being fired at Israel.[26] Laser designation can also be achieved by the aircraft delivering the ordnance. In such circumstances, the aircraft will "paint" the target with a laser and drop its ordnance, which will then ride the laser down to the target.[27] This maneuver requires the aircraft to be put in harm's way, risking the life of the pilot. Aircraft can employ laser-guided munitions quickly to maximize the chances of taking out a target of convenience or opportunity. They are, however, not the backbone of precision-guided munitions for the U.S. armed forces today and will not be in the future.

By 2003 the U.S. military had moved primarily to munitions that used guidance from the global positioning system (GPS).[28] Using GPS to obtain its location vis-à-vis the target's location, the preprogrammed tail package directs the munitions to the target.[29] There are inherent advantages to using GPS-guided munitions over laser-guided munitions, such as the former's all-weather capability, including in fog, rain, and dust storms. In past years, the programming of GPS-guided munitions tended to be a laborious process that did not easily allow the munitions to be used on targets of opportunity. It was the case until the U.S. military's intervention in Afghanistan following the September 11, 2001, attacks (9/11) in New York City and Washington DC. However, after 2001 the ability of ground personnel or personnel aboard the aircraft to upload information directly into the warhead allowed faster targeting of GPS-guided munitions in direct support of U.S. military personnel and ground forces.[30] By 2003 the use of GPS-guided munitions became commonplace and was truly an asset to U.S. military personnel fighting against the Iraqi Army.

GPS-guided munitions posed a problem for U.S. forces. The small diameter bomb was being developed but was not yet in production. There have been some novel approaches to lowering the payload of air-delivered ordnance, such as the five-hundred-pound bomb, by reducing or eliminating the actual explosive in the bomb to allow a smaller detonation and reduce collateral damage.[31] However, this option was not available prior to 2008.[32] Consequently, in Iraq for five years the smallest munitions available with the GPS-guided tail package used the five-hundred-pound bomb. This ordnance creates a large blast and massive collateral damage when used in built-up areas. Consequently, the decision to use such a device had to be well thought out.

While employing a GPS-guided weapon against troops in the open would be to the advantage of U.S. military personnel, troops in the open are not commonplace, especially in a counterinsurgency environment. At the beginning of the 2003 intervention in Iraq, Iraqi forces were clearly identified on the battlefield, and GPS-guided weapons were extremely effective. Using these munitions to take down telecommunication centers, large buildings, transportation hubs, and bridges proved quite impressive in crippling Iraq's command-and-control facilities, which organized resistance to U.S. military intervention.[33] A high-yield precision munition is a technically capable widget, or device, that allows for the delivery of ordnance to precise points in space and time. Unfortunately because of the large field of the blast, it is best used against military targets in the open, hardened targets such as reinforced concrete, or large static structures such as communications hubs. Large and powerful air-delivered ordnance represents a danger not only to the civilian population and civil structures but also to U.S. military personnel who may be in the proximity of the blast.

The effectiveness of GPS-guided and laser-guided munitions in a force-on-force confrontation is beyond question, but their actual use in a counterinsurgency environment can, in fact, be counterproductive. The high yields and devastating blasts in populated areas will inevitably lead to casualties and the destruction of civilian infrastructure. Such concerns may not be in the forethought of a military commander whose primary mission is a conventional objective or conventional forces. A military commander may see the inevitable destruction of life and property simply as collateral damage, a nuisance, or an unfortunate occurrence. In commanders' minds the military objective will have primacy over keeping the civilian population from harm.[34] It is precisely this type of thinking that becomes very dangerous when using high-yield ordnance in sizes from five hundred to five thousand pounds in a counterinsurgency environment.

Precision munitions, with a very high probability of landing directly on the target or within three meters of it, do not alter the fact that the devastation they wreak in a civilian area will inevitably harm a counterinsurgency effort there.[35] This observation in no way implies that using these weapons in built-up or civilian areas should be regarded as illegal or subject to constraints in a counterinsurgency effort. Every counterinsurgency conflict is different; the strategies employing these tactics all have to be adapted to the actual counterinsurgency in the milieu in which it is being fought. However, such weapons systems must be very carefully managed in order to achieve the goals of stabilizing the host nation or destroying the extremist elements. For the most part, the United States did it quite well in Iraq.

The employment of high-yield precision-guided munitions does have a utility that goes beyond the battlefield. The weapons can, and often do, impress both insurgents and local populations as being of profound destructive power. In 2008 the Fourth Brigade, Second Infantry Division, responsible for operations in Diyala Province in Iraq, expended more ordnance than any other brigade in the country did.[36] The irony of this situation is that the brigade continued to expend approximately the same amount of ordnance on a monthly basis after it reduced violence in Diyala Province by more than 90 percent.[37] The express purpose of this mission was to keep the local population in awe of U.S. military firepower.[38] While in the village of Abu Khamis, south of Baquba in Diyala Province, the commander of Alpha Battery, Twelfth Field Artillery, attached to the brigade, pointed out a petrol station that had been destroyed by a joint direct attack munition (JDAM). The petrol station had been rigged with an improvised explosive device (IED), which his soldiers had found while sweeping through the village. While the explosive ordnance disposal unit was fully capable of disarming and removing the IED, the unit made the decision to use a JDAM on the petrol station and completely destroy both the device and the petrol station. The true purpose of the action was to unequivocally impress upon the local population of Abu Khamis the destructive firepower at the U.S. military's disposal.[39] Such an action may be thought of as counterproductive in a counterinsurgency effort, and if used to the extreme, it certainly can be. However, while always a subjective judgment call, such an action can also be very effective in impressing upon the local population the seriousness of the military unit operating in the area. For people unfamiliar with explosive blasts, and even people familiar with blasts on a small scale, the detonation of five hundred to five thousand pounds of high explosive has a truly dramatic sound and feel. In this particular case, civilian infrastructure was purposely destroyed with nothing but the most flimsy veneer of acquiescence to military protocol—that is, the destruction of an improvised explosive device. However, it proved to be a well-considered and purposeful action that seemed successful, as several days later locals pointed out other IEDs to American personnel.[40] It should also be noted that Iraqi civilians in the area did have other means to obtain petrol.[41]

The IED example is meant to illustrate that the technology used in the 2003 intervention in Iraq could be used in several different ways. Given the firepower-centric nature of the U.S. military and its conventional mind-set, coupled with no sound counterinsurgency practice between 2003 and

2006, the use of overwhelming power led to the alienation of the Iraqi people rather than their cooperation.[42] However, those same weapons systems and technical capabilities are not inherently counterproductive to a counterinsurgency effort. Their employment requires a great deal of thought and in-depth knowledge not only about the weapons systems, the tactics employed, and their limitations but also of the culture in which they are going to be used. It is critically important for a junior officer to assess and understand the reaction of the population to such a weapons system, in this case a five-hundred-pound JDAM, when he is judging whether to launch the weapon. When the junior officer employs technology and the killing power at his disposal, that officer must understand the full spectrum of consequences for the counterinsurgency operations in which the unit is involved. Indications are that in Iraq from 2003 to 2006 the use of such firepower tended to be detached from a sound counterinsurgency effort.[43] The employment of firepower during this time tended to focus on force protection and overwhelming force.

Another profound technical advantage for U.S. military personnel in the 2003 intervention in Iraq, as opposed to the U.S. military's 1991 intervention in the state of Kuwait, was a dramatic increase in the effectiveness of body armor. Typical body armor in 1991 consisted of a Kevlar vest, which was generally a multilayered Kevlar vest modeled after the old nylon flak vest that U.S. forces wore in Vietnam.[44] The idea behind the vest was to create a barrier around the torso that was capable of stopping grenade fragments and other small objects from penetrating the chest cavity and causing dramatic internal injuries. However, the Kevlar vests in 1991 were not designed to, nor were they capable of, stopping a high-powered round from a rifle such as an AK-47.[45] While the vest from 1991 was capable of slowing down a round and lessening the injury sustained by a soldier, the damage to the chest or internal organs still could be quite severe. By 2001 special operations units were routinely wearing body armor with ceramic plating that was capable of stopping or deflecting high-velocity bullets.[46] By 2008 ceramic level 4 protection plates worn in the front and back of body armor, and in some models on the sides, were capable of stopping up to six rounds from an AK-47 and even a high-velocity steel core sniper round from a 7.62mm rifle.[47]

The Kevlar helmets typically worn by U.S. military personnel in 1991 were only capable of deflecting a rifle round from an AK-47; any direct impact from a round to a Kevlar helmet would result in almost certain death or grave injury. By 2008, the modular integrated communications helmets (MICH) were capable of stopping not only a direct impact from a 7.62mm rifle round but also high-velocity 5.56mm rifle rounds.[48] In one specific case a soldier in C Troop, Second Squadron, First Cavalry Regiment, attached to the Fourth Brigade, Second Infantry Division, was struck in the back of the head by a high-velocity 5.56mm round. The round penetrated the majority of the layers of Kevlar in the MICH before being deflected upward, traveling between the layers, and stopping in the crown of the helmet. The soldier described feeling as though he had been struck in the back of the head with a baseball bat. His only resulting injury was a headache.[49]

The implications of this protection for individual soldiers should be self-evident in a conventional environment of force-on-force confrontation between two armed groups. Most American soldiers serving in Iraq in 2008 clearly understood that being shot was not necessarily synonymous with a life-threatening injury. In fact soldiers commonly said that the body armor provided gave troops the confidence in a combat situation to survive contact with enemy forces.[50] This level of protection and confidence directly equates to higher morale and better unit cohesion in firefights. While soldiers' extremities are still prone to injury, which can lead to severe disability or possibly even death, for the most part rifle fire and grenade fragments have been almost eliminated as causes for combat fatalities. Of course combat fatalities actually do occur from small arms fire, and several scenarios demonstrate how small arms fire is capable of killing American soldiers in firefights. A round impacting low in the pelvis and traveling upward, impacting into the face or directly under the armpit, then traveling laterally through the chest cavity can still create catastrophic wounds that result in the fatality of a soldier or, at a minimum, leaving him or her in a critical state.[51] At Forward Operating Base (FOB) Warhorse, just north of Baquba, the Fallen Raiders Wall in the brigade's tactical operations center had a picture and the cause of death of every soldier killed from Fourth Brigade, Second Infantry Division. Of the fifty soldiers who were killed between September 2007 and April 2008, only four soldiers—less than 10 percent of the brigade's overall casualties—died due to small arms fire, typically rounds fired from AK-47s.[52] This number does not, however, include soldiers who died from sniper fire. A well-placed sniper round will usually be targeted at more vulnerable areas of a soldier and, therefore, be more likely to penetrate the protective measures worn by U.S. troops in Iraq.

Protective measures used in Iraq to keep soldiers safe were not limited only to personal protective equipment. A number of vehicles were developed and fielded to protect soldiers from the most lethal threat they faced in Iraq, improvised explosive devices. Of the fifty soldiers on the Fallen Raiders Wall, IEDs accounted for twenty-five of the fatalities. Six of the twenty-five were the result of a single explosive device placed in the interior of a house that detonated and killed an entire squad. The remaining nineteen fatalities were the result of roadside IEDs; these bombs usually targeted vehicles, though occasionally foot patrols were attacked. With the high proportion of fatalities resulting from IEDs, it became clear that combat troops in Iraq required better-protected vehicles than what was available when the intervention began in 2003, much less in 1991. Commonly in 1991, military vehicles such as the high-mobility multipurpose wheeled vehicle, or the Humvee, did not even have metal doors; instead, they had canvas doors that zippered shut.[53] Such vehicles are entirely suitable for a conventional force-on-force engagement in which transport and utility vehicles such as the Humvee are normally not expected to come under enemy fire. Typically conventional engagements occur between armored forces with infantry support, and Humvees and other light-skinned vehicles would normally be to the rear or at least working in concert with combat vehicles that are capable of delivering firepower at range and protecting them. Patrols in Iraq, however, required vehicles that offer mobility and speed. In such situations, insurgents would inevitably target soft-skinned vehicles and the military personnel riding in them commensurate with sound insurgency tactics.[54]

Today's Humvees are markedly different from those used in Iraq in 2003 and in the 1991 Gulf War. The Humvees are up-armored with metal doors and steel plating to protect the soldiers inside, as well as turrets to provide additional firepower. Some of these turrets are enclosed in armor, but others simply have armor plating around them. Typically the Humvee with its added protective measures now weighs twelve thousand pounds, or more than twice what the standard Humvee weighed in 1991.[55] These vehicles are still vulnerable to large IEDs but maintain the ability to move quickly over a variety of terrains and in different capacities. The requirement to up-armor Humvees again demonstrates a lack of foresight and planning by the U.S. military to accurately predict and understand threats that U.S. military personnel would face in future conflicts. The overwhelming focus on conventional force-on-force engagements did not take into account the requirement that soldiers in transport vehicles had to be protected from small arms fire and small IEDs, as well as the rocket-propelled grenades that insurgents typically use.[56] The changes made to the Humvee demonstrate the capability of the U.S. military to adapt to the circumstances in which it finds itself operating. That adaptation is a saving grace for the U.S. military, but lessons learned are bought with the blood of soldiers, an unfortunate result of substandard prior planning and budget allocation.

Other vehicles, such as the Stryker combat vehicle and the mine-resistant ambush-protected (MRAP) vehicle, provide soldiers with added protection and options for firepower.[57] These vehicles are generally impervious to small arms fire, heavy machine guns, grenades, and smaller IEDs, with the MRAP providing outstanding protection even against midsize to large IEDs.[58] These vehicles are costly both in terms of dollars and in political capital as budgetary allocations are finite; however, they are specifically designed to enhance the survivability of soldiers while also allowing them more tools with which to accomplish their objectives. While no vehicle ever will be totally impervious to all IEDs, especially some of the more advanced explosive penetrators, they offer a level of protection that has markedly enhanced soldier survivability.

Technical advances in combat vehicles have included additional protective measures for individual soldiers. The blue force tracking system embedded in military vehicles today allows soldiers to determine not only their own location but also that of other friendly combat vehicles in the area.[59] In essence the system works with a video screen that shows a real-time, live map of the surrounding area with the vehicle in the center and blue icons indicating the location of other friendly vehicles nearby.[60] This system greatly enhances situational awareness for vehicle commanders and operators. Because they can see where friendly forces are located, the likelihood of friendly fire is reduced. The system also enables a commander to position his forces with a clear understanding of where joining forces are located. Such situational awareness is critically important to ensure areas of responsibility remain separate between different units to prevent blue-on-blue incidents.

Other advances, such as the one system remote video terminal, allow real-time feed from unmanned aerial vehicles (UAVs) flying over the battle space to be sent directly to a commander's vehicle, enhancing the ground commander's situational awareness of the mission.[61] While such measures may be far from the full-spectrum awareness envisioned by technophiles who speak enthusiastically of the revolution in military affairs, the application of these capabilities provided ground commanders in Iraq with an overwhelming advantage in situational awareness and command and control.[62]

Unmanned aerial vehicles were extensively employed in Iraq. Their use by the U.S. military increased dramatically since the invasion and remained in high demand. Such technical capabilities offer the U.S. military an ability to observe, track, and target insurgent groups or individuals while exposing U.S. military personnel to little danger. While the technology behind UAVs has been evolving for many years, it was not until after 9/11 that such technology has come to the forefront of military capabilities.

It is hard to overstate the effectiveness of UAVs in a war-fighting capacity. In the conventional realm of warfare, they offer situational awareness to main force units by allowing the tracking and targeting of enemy formations, as in the initial invasion of Iraq in 2003. This ability to locate, track, and target enemy formations provides U.S. military forces with an obvious technical edge in the capability to engage, and be effective, in force-on-force scenarios. While there is great speculation about the actual capability of such aircraft in the realm of intelligence gathering and surveillance, and as strike platforms, much of their actual capability remains classified.[63]

The coordination and control of UAVs depends on the particular type of vehicle and its capabilities. Some UAVs are the assets of corps commanders, division commanders, brigade commanders, and even as far down the chain of command as battalion and company commanders.[64] The U.S. military would like to allow commanders at the lowest level possible to utilize UAVs for tactical operations; however, some vehicles are not designed for directly supporting tactical operations.[65] Vehicles such as the Global Hawk UAV further command and control through increased situational awareness of higher-echelon commanders and support their missions or objectives.[66] Smaller UAVs, such as the Shadow, are reserved for brigade commanders but in turn are tasked to support battalions within the brigade's area of operations. The spreading of these unmanned aerial vehicle assets to different echelons of command is a natural development of the programs as they move to best utilize technical capabilities in support of ground commanders.

In Iraq in the spring of 2008, during combat patrols with the Fourth Brigade, Second Infantry Division, in Diyala Province, this author spoke with several commanders about their opinions and perceptions of the value of UAVs. The overwhelming response was that the commanders wanted more of these assets, which would allow greater coverage of the area of responsibility and enhance command and control over the troops in the field.[67] It is important to note that besides surveillance, UAVs also provide utility functions such as relaying radio messages between ground force commanders in the field and command-and-control teams at the battalion or brigade tactical operations center, which oversees and at times directs operations from the rear-echelon area.[68] The augmentation of communications is, and will remain, a vital role for UAVs and will continue to make them a much desired asset for commands at every level.

With a large area of responsibility, the surveillance of enemy forces and their movements becomes imperative for the brigade commander. Every asset that does not risk the lives of American soldiers or force a brigade commander to commit badly needed troops elsewhere immeasurably enhances that commander's ability to effectively control and plan operations within his area of responsibility. To effectively support the brigade commander, the four Shadow unmanned aerial vehicles, which were assigned as brigade assets, were required to fly missions twenty-four hours a day, seven days a week, but the Shadow UAVs are actually designed to fly only twelve hours a day. To meet the demands of combat operations in situational awareness, the twenty-three-man platoon under Capt. David Siler of Headquarters and Headquarters Company Troop, First Cavalry, Fourth Brigade, Second Infantry Division, had to devise a way to keep the UAVs in the air twenty-four hours a day. Siler accomplished it over the course of the fourteen-month deployment while also being directly responsible for more than three hundred insurgents killed at the loss of none of his own men.[69] The need to double the normal operational tempo of the Shadow UAV simply illustrates the overwhelming desire of commanders for more unmanned aerial vehicle assets. While the brigade commander is fully capable of requesting and receiving specific UAV support from higher echelons, such as Predator unmanned aerial vehicles from the division level, these assets are in high demand and may not always be available to individual brigade commands.[70]

An illustration of the technical capabilities of an unmanned aerial vehicle and the opportunities it can provide in a counterinsurgency effort was very clearly revealed on a night air assault mission in early April 2008. Twenty-two American soldiers were standing by at the heliport in FOB Warhorse outside Baquba for a mission of opportunity that night. A UAV flying just east of the town of Khalis spotted three insurgents placing an IED in a culvert by a road. The controllers of the Shadow UAV relayed the individuals' coordinates to an air weapons team standing by, in this case, an AH-64 Apache attack helicopter. While the helicopter was en route, two of the insurgents left the third behind and drove toward Khalis on a motorcycle. The air weapons team eliminated the third insurgent as he was walking down the road, while the Shadow UAV continued to follow the motorcycle and watched it as it weaved through the streets of Khalis.[71] At the same time, the position of the IED was relayed back to the battalion tactical operations center for subsequent disarming or destruction by combat engineers the next day.[72]

The two insurgents on the motorcycle eventually parked it and mingled with the crowds in the town of Khalis. It was decided that destroying the motorcycle or attempting to target the insurgents would be counterproductive as all three were situated in a densely populated residential area. The Shadow UAV maintained surveillance for more than two hours until the insurgents returned to the motorcycle and drove east out of Khalis.

It then followed the insurgents to a chicken coop roughly ten miles east of Khalis and relayed the coordinates to the battalion tactical operations center. Within minutes the twenty-two American soldiers and a civilian observer loaded up on two Black Hawk helicopters and flew ten minutes to the chicken coop at which the insurgents had stopped. While in flight the Shadow UAV continued to surveil the chicken coop and was prepared to follow any individuals who left. Within ten minutes of the helicopters' landing, the soldiers had the chicken coop under control and had captured four insurgents.[73] One of the captured insurgents was a corrupt police officer who was wanted in another province of Iraq.[74] Two of the insurgents admitted to being two of the three individuals placing IEDs that night, as had been observed by the unmanned aerial vehicle, and were subsequently turned over to the Iraqi authorities for prosecution and imprisonment. The fourth insurgent was the most wanted target in Diyala Province; he commanded extremist Shia militias that were involved in sectarian violence against Sunnis and attacks on U.S. forces.[75]

In a conventional conflict, the apprehension of four individuals usually would be of little consequence to an overall military effort or campaign. However, within the realm of a counterinsurgency, the ability to effectively target and apprehend individuals responsible for insurgency activity, especially ones in leadership positions, reflects sound doctrine and results.[76]

This example has many enlightening facets that speak directly to the role of technology in the counterinsurgency effort in Iraq. First and foremost the effective apprehension of the four individual insurgents, two of whom were considered high-value targets, was the result of a UAV and its capability for remote surveillance. In this case the targets were identified without ever putting a single American soldier at risk. The ability of the Shadow UAV to operate for prolonged periods with all-weather surveillance capability allowed it to track the specific motorcycle used by the insurgents. It also enabled commanders back at the tactical operations center to determine the potential scale of collateral damage that would be caused if they engaged the motorcyclist targets within the built-up area of Khalis. The commanders would have been within their rights to target the individuals, even in a built-up area, if a strike had been deemed proportional.[77] However, by combining the technical capabilities of the

Shadow UAV with a sound appreciation of counterinsurgency doctrine, they clearly understood that they could be patient and wait for the insurgents to move to a more accessible area for the air assault force.

Once the targets reached the chicken coop, the technical capabilities of air assault, even in the middle of the night, ensured the rapid deployment of American troops to apprehend the insurgents in a timely manner before they could escape.[78] The body armor worn by the twenty-two soldiers offered enough protection that no likelihood existed of the enemy insurgent force easily wounding and overwhelming the U.S. infantry involved in the operation. To augment the ground force, the two Black Hawk helicopters used as transport lifted off after the insertion and provided air support at a very low level above the ground. Circling above the Black Hawk transports, with their machine guns, were the air weapons teams of two Kiowa helicopters. Above the Kiowa helicopters, the original Shadow UAV provided continuous surveillance of the target area and relayed communications from the inserted ground force back to the tactical operations center. Above the Shadow UAV flew an F-16 with a JDAM capable of leveling the chicken coop if the resistance was too fierce.[79] Brought together at very short notice, all of these elements worked together, in concert and in the middle of the night, to apprehend a target of opportunity in a manner that reflected sound counterinsurgency doctrine. This event was typical of the well-executed counterinsurgency strategy of the Fourth Brigade, Second Infantry Division, in Diyala Province.

This mission clearly demonstrates the potential effect that UAVs can have for conducting successful operations. It is not hard to understand why ground commanders in Iraq have such a strong desire to obtain as much information as possible and why they want to have as many UAV assets as possible. As wonderful as such technical capabilities may be, however, it should be noted that only four insurgents were captured while the population of Diyala Province is more than a million people and undoubtedly contains many more actual or potential insurgents.

III. Diyala Province, 2008

The Diyala Province of Iraq is roughly the size of Maryland. It is a very large area and population for a single combat brigade to try to manage in a counterinsurgency role. The Fourth Brigade, Second Infantry Division, was tasked with engaging and eliminating al-Qaeda's presence and influence in the province.[80] Such a role for a little more than five thousand soldiers is daunting. Even with so few troops, the brigade had to undertake its mission. Col. John Lehr, the commander, wisely chose first to gain control of population centers and then to separate insurgent groups and their activities from the general urban population as a whole. He recognized that a counterinsurgency effort would be won or lost through the involvement of leaders of the indigenous populations of Diyala and their ability to minimize violence. This recognition resulted in a sound counterinsurgency strategy that focused on the local population rather than solely on combat operations in suspected insurgent base camp areas.[81] In 2007 military leaders understood that many of the population centers in Diyala Province were held by insurgents who were based in the cities and had fallback positions in more rural areas. A conscious effort was made to move elements of the brigade into the population centers first and to remove the insurgents from direct contact with the population while ignoring those rural centers that were possible base camps. It was a logical plan given the limited troop numbers available to accomplish such a mission in a large province.

In 2005 under the name the Islamic State of Iraq, al-Qaeda in Iraq (AQI) set up a shadow government based in Baquba, the provincial capital of Diyala Province.[82] The group chose the name in an attempt to rebrand AQI as a different entity with a legitimate Islamic mandate to rule Iraq. U.S. forces in Baquba largely ignored AQI because they lacked the manpower to confront the insurgent forces in the city while continuing to maintain operations in Al-Anbar Province to the west. However, the constant stream of suicide bombers and vehicle-borne improvised explosive devices trickling into Baghdad, only thirty-five miles away, created a political imperative that the city of Baquba had to be captured from the insurgents.[83] Operation Arrowhead Ripper managed to recapture the city block by block during mid-2007.[84] The city was not fully retaken, as pockets of insurgents would return occasionally through late 2007, but for the most part the insurgents had been forced out of Baquba to outlying areas.

The Fourth Brigade, Second Infantry, deployed to Diyala Province in the spring of 2007 and took part in Operation Arrowhead Ripper to regain control of Baquba, which was centrally located in the most populated area of the province. Most of the population of Diyala Province lives in a Y-shaped area, with Baquba at the junction of the Y. The populations extend south-southwest from Baquba toward Baghdad and Khan Bani Saad, northeast through the Diyala River Valley toward Al-Muqdadiyah, and northwest past Khalis toward Balad. Commonly referred to at FOB Warhorse as the Y, the area represented the majority of the population that had to be secured from insurgents.[85] Also many small villages were scattered throughout the province in rural areas, but these communities' populations usually numbered in the hundreds.

Throughout 2007, the brigade pushed out from Baquba incrementally in the three directions of the Y to bring an ever-increasing area under its control.[86] Colonel Lehr attempted to take control of the largest population centers and flush insurgents out into the surrounding countryside, where they could receive less material support. As U.S. troops took control of areas, they raised and financed local militias to provide security in their communities. This program, instituted theaterwide by Gen. David Petraeus, was known as Awakening Councils in Arab circles and as Concerned Local Citizens (CLCs) by the coalition forces.[87] As he lacked available manpower, Colonel Lehr had to raise local forces to control areas that his units had already cleared.[88] These councils allowed the U.S. units to continue to move farther into insurgent-controlled areas and secure more of the populace.

Diyala's populace presented further challenges, as the province and the Y contained a convoluted mixture of Sunni and Shia areas. Often one village would be Sunni and the next would be Shia. In the case of Big Barwanah and Little Barwanah, just east of Al-Muqdadiyah, the Shia of Little Barwanah and the Sunnis of Big Barwanah exchanged nightly gunfire and occasional mortars and rocket-propelled grenades across the ten-meter-wide Diyala River.[89] North of Khalis in 2006 Sunni and Shia militias held company-size firefights, and Khan Bani Saad was a block-by-block mixture of Shia and Sunni.[90] After the sectarian violence that had occurred in 2005–6, it was extremely difficult for Colonel Lehr to institute neighborhood watch–style programs without having neighborhoods going to war with other neighborhoods or a neighborhood going to war with itself, as often was the case in East Baquba.[91] Different Sunni militias, such as the 1920s Revolution Brigades, were at odds with other Sunni militias in the same parts of East Baquba.[92]

The peace was kept by assigning each militia its own territory, which the militia would retain as long as it kept al-Qaeda out and did not start any serious violence with neighboring militias. In the event that violence occurred, the militias all knew that the brigade would arrive to restore order and that they might face a possible firefight against the overwhelming technical advantage of a Stryker brigade combat team. This situation created a fragile truce backed by fear and then allowed a cascading effect, lending itself to good counterinsurgency practice. Colonel Lehr's men were able to move outward and free more people from insurgent control. The increased security allowed for reconstruction projects, which improved the infrastructure and led to a resumption of the economy and therefore a happier populace.[93] In short, this effort was good counterinsurgency strategy in practice.

However, the massive area of Diyala Province—both in the area controlled by Fourth Brigade, Second Infantry Division, and the rural areas controlled by the insurgents—made effective implementation of sound counterinsurgency practice difficult at best. Any flaring of sectarian tensions could cause U.S. troops to stop expanding their area of influence in order to quell the problems.[94] As the territory under the influence of the host nation expands without increased troop numbers, the thinner and more tenuous the military's control will be. The requirement for ever-increasing resources puts a constant strain on the counterinsurgency effort.

The counterinsurgency role of the brigade in Diyala Province extended beyond kinetic operations, such as Arrowhead Ripper, and the counterinsurgency practice of forming local militias. Civil affairs played a large part in the effort to win the people's allegiance to the nascent Iraqi government. The role of civil services was well understood throughout the brigade. When asked what was the biggest challenge the brigade faced, Capt. Phillip Mundweil replied, "We have to get services up and running or the people will rightly demand to know why we are here. Just [providing] security is not enough."[95] It may sound fairly simple to get basic services up and running to restore people's faith in governance and draw them away from an insurgent ideology such as that of Al-Qaeda's or other groups espousing Salafist tradition.[96] But the reality is very different when operating in a foreign culture that is fundamentally different from that of the United States. When asked about the cultural challenges, Mundwiel observed: "It's Iraq. They were socialist under Saddam and everything was allotted to them in a central way. They don't know how to communicate between departments and they don't take initiative. We kind of have to show them how or who to talk to. After that they work pretty well. It is so new to them but they are starting to get it."[97]

The culture itself presents obstacles, such as corruption, in other ways. Pervasive corruption presents political opponents of a counterinsurgency effort with an issue they can exploit. The insurgent force, however, has to make a judgment call on what constitutes too much corruption. As Mundwiel notes: "It's [corruption] there. Normal corruption. It's a different society and it's endemic. If someone tries charging me $10 for a window that costs $6, I am going to say no way. But if they charge me $7 and pocket one, well, that's the cost of doing business. A big problem is no accountability."[98]

The brigade's experience in Diyala Province demonstrates what is possible with today's technical capabilities when linked with sound counterinsurgency practice. But it also highlights many of the challenges that a force faces when trying to operate with fewer than the optimal number of soldiers. In mid-2008 the brigade was due to rotate back to the United States, but it had been continuing to push farther into insurgent territory and bring more Iraqi policemen, Iraqi Army personnel, and Awakening Councils/CLCs up to standard to help in securing the populace.[99] The Fourth Brigade, Second Infantry Division, in Diyala Province clearly showed that the U.S. military is capable of learning and adapting to face counterinsurgency challenges. It also demonstrated the importance of technical tools in counterinsurgency efforts. Only time will tell if these efforts are enough for that specific region with its unique variables.

IV. The Limits of Technology in Iraq

There were clearly two phases, at least, in the conflict in Iraq from 2003 to late 2009. Of course, one should recognize that more phases are possible in the future and that no amount of speculation will prove as accurate as hindsight. A convenient classification into the conventional phase and the counterinsurgency phase is most apt. Without question the initial invasion of Iraq in 2003 by U.S. military conventional units using standard force-on-force doctrine against the Iraqi Army of Saddam Hussein was a classic conventional conflict to establish control of major population areas and destroy the enemy's capability to wage war. As U.S. forces massed and moved north into Iraq, resistance was sporadic. The outcome, while predictable, was swifter even than had been expected by American commanders.

During the conventional phase of the invasion in Iraq, technology had few limiting factors. Most U.S. systems performed as expected and provided overwhelming firepower to U.S. ground forces. Even in adverse conditions, such as dust storms, the ability of U.S. military hardware to target and destroy Iraqi positions, vehicles, and troops was absolutely one sided. U.S. forces performed as they had in the 1991 Gulf War, which was a fairly accurate prediction for the subsequent engagement of U.S. forces against a degraded Iraqi military in 2003. The overwhelming superiority of U.S. technical hardware and targeting ability directly resulted in the extremely low American casualties on the way to Baghdad, and the overwhelming technical ability and firepower associated with U.S. forces were not lost on the Iraqi population or its military. Most Iraqi forces chose not to fight the American troops, as the outcome of any such battle was already a foregone conclusion. The Americans encountered some exceptions, with the battle for Objective Peach resulting in the annihilation of an Iraqi brigade by a single U.S. company, which itself took almost no casualties.[100]

Had this conflict theoretically occurred in 1945 perhaps the outcome would have been different. Perhaps the defeated armies of Iraq would have surrendered, and the government of Iraq would have capitulated and brought the conflict to an end. But it was clearly not the case in Iraq in 2003. Despite the overwhelming firepower and military capability of U.S. forces, their stated objective was not the destruction of the Iraqi Army but the removal of Saddam Hussein from power. Because of the inherent nature of the power structure in Iraq itself, the stated objective meant removing not only Saddam Hussein but also those who had benefited from his rule. Therefore, those who had benefited—namely, the Sunnis and the Ba'ath Party members—found themselves at a disadvantage under the new regime and as a whole were subordinated to the Shia majority.[101] If Western policy makers who were involved in the intervention clearly appreciated this fact, then they should have seen it as a very likely catalyst to an insurgency against what was certainly going to be seen as a foreign occupation of a Muslim land. In short all the makings for a Sunni and likely fundamentalist Muslim insurgency were at hand.

The second phase of the U.S. military intervention in Iraq started in 2004–5 when insurgents stepped up attacks against U.S. forces, as well as nascent Iraqi security forces, and attempted to destabilize the host nation, pursue political objectives of indigenous groups, and draw the United States into a protracted conflict with fundamentalist Muslim groups, most notably al-Qaeda. Clearly this phase is juxtaposed to the first phase of the conventional conflict between U.S. forces and Iraqi conventional forces. It is puzzling that the U.S. military did not seem prepared to engage in a serious counterinsurgency effort at the time. There are several possible reasons for this, none of which is terribly flattering to the U.S. military or to the civilian oversight by the Bush administration.

It is possible that many military professionals were simply incapable of coming to grips with a counterinsurgency effort that differed so widely from their own definitions of their roles as war fighters in a conventional army that had been focused on conventional threats. In essence these professionals would have to learn a new style of warfare, one that is far more uncomfortable both in its prospects for success and in its demands for creative thinking by the individuals involved. To say that the U.S. military is incapable of thinking about counterinsurgency warfare would be unfair. In fact there has always been a community in the military, often called snake eaters in an obscure reference to U.S. Army Special Forces, that has been more unconventionally oriented. It is equally possible that, as John Nagl argues, the U.S. military as an institution is not readily able to change or adjust its thinking and strategies due to a rigid conception of itself.[102]

This problem may partially originate from the early 1990s and stem from the guidance of the Defense Department's Office of Net Assessment under the direction of Andrew Marshall. Net Assessment does not drive and dictate strategy or doctrine, but it does have a small influence. With the fall of the Soviet Union, leaving the United States as the only super-power in the world, Marshall became convinced that the next threat to U.S. national security would be a rising and powerful peer rival in China.[103] Andrew Marshall may be forgiven for making this judgment, which has subsequently not yet materialized. He has headed the Office of Net Assessment since its creation in 1973, and the military's emphasis in the first half of his tenure was on countering the threat from the con-ventional and nuclear forces of the Soviet Union.[104] It may have been a far-fetched idea then to see that the future conflicts and threats to U.S. national security would be as likely to come from non-state actors as from state actors, and far be it for anyone to criticize an individual's inability to accurately predict the future. However, if the Office of Net Assessment, in fact, had advised the United States to focus on non-state actors and terrorist threats, then the U.S. military might have been in a slightly better position to handle the switch from conventional warfare to unconventional warfare in Iraq following the 2003 invasion. Of course, it is more than possible that the bureaucracy and the military would have resisted such an emphasis had it been proposed.

Regardless of its origin, whether it was a combination of factors or a single factor, the military's inability to move quickly from a conventional mind-set and strategy to an unconventional mind-set and strategy serves only as a lesson for the future. What is of prime importance is that the U.S. military spent two years in Iraq focused on force protection and tactical war-fighting capability rather than shifting its mind-set to a nontechnical mode of warfare by counterinsurgency forces, which rely on intelligence rather than on firepower.[105]

Arguably some of the most important factors for the rise of the insur-gency in Iraq in 2004–5 were the policy decisions the Coalition Provisional Authority made during that time. Clearly no technical capability, war fighting or otherwise, is going to provide relief from bad policy decisions made by politicians. Policies by, or pertaining to, the host nation in an effort to govern its people and society in a legitimate manner are not tech-nically driven. In fact bad policy can easily hamper a counterinsurgency effort more thoroughly than any technical shortcoming by the counter-insurgency forces on the battlefield.

One of the most notable failures on the part of the Coalition Provisional Authority was Paul Bremer's decision to disband the Iraqi military.[106] The Iraqi armed forces had been a stabilizing institution in Iraq and was argu-ably the only effective organization for maintaining control of civil society, especially in light of the limited number of coalition troops used in the intervention. Bremer's decision to disband the Iraqi military left tens of thousands of Iraqis disenfranchised and unemployed. That many of these men had wives and children to feed presented a huge recruitment oppor-tunity for insurgent groups bent on hampering the coalition's efforts to stabilize Iraq and build a legitimate government. As a practical example, if an insurgent, whatever the affiliation, approached a soldier from the former Iraqi military who could no longer provide for his family and offered to pay him $50 or $100 to plant an improvised explosive device, then what possible option would that individual truly have?[107] The ex-soldier might not have been sympathetic to the insurgency, but he was very likely unsympathetic to the Coalition Provisional Authority, which had cost him his job, his ability to feed his family, and therefore his dignity. The man would prob-ably do whatever he had to do in order to survive. In 2004 there simply was no technical solution to dealing with 200,000 unemployed and angry Iraqi men who had the duty to provide for their dependents.

Between 2005 and 2007, as sectarian violence exploded and the various insurgencies gained momentum, the perception that technology alone drives capability was demonstrated to be spectacularly wrong. At the time, the prevailing opinion was that the Iraqis would have to fight their own war and that it was pointless to expect U.S. forces to fight it for them. This attitude is, in essence, correct; host nation forces must be capable of eventually confronting and defeating insurgents. A common argument from the Vietnam era was that the United States tried to fight a war for the Vietnamese who should have been fighting it themselves. However, between 2005 and 2007 the U.S. strategy in Iraq could not have been worse as it directly contradicted sound principles of counterinsurgency warfare.

In Iraq, American leaders assumed that the U.S. military's presence in cities and towns would alienate the populace by giving the impression that the U.S. forces were there to occupy the country rather than to liberate the people of Iraq.[108] Therefore, U.S. forces retreated to large bases and encouraged the Iraqi police and new security forces to take control of civilian areas.[109] The net effect of this strategy, or rather lack of strategy, was to leave the population vulnerable to insurgent groups and criminal organizations without any alternative for their safety. Added to the mix were the dismissed Ba'athist officials and the unemployed former members of the Iraqi Army. The inevitable result was an insurgency that ballooned out of control; that had an almost endless supply of recruits, who had no other viable alternatives; and that faced a U.S. force posture that lent itself to endless ambushes when units left their bases. These factors combined for a spectacular counterinsurgency failure.

The environment in 2005–7 in Iraq, where the U.S. military refused to move into populated areas to secure the populace, negated the technical advantages of firepower. Instead of pursuing a proactive approach to counterinsurgency warfare, the U.S. military was risk averse, overstressing force protection and the minimization of casualties. By allowing insurgent elements free rein to recruit members and to construct improvised explosive devices, U.S. forces guaranteed that IED attacks and small arms ambushes would occur on a regular basis when U.S. forces did try to patrol.[110] Concentrating on force protection by holing up in bases, U.S. forces cut themselves off from meaningful cooperation with the local population, including the gathering of human intelligence. Thus, in turn, they were less able to protect themselves in the long run. It also had the negative effect of leaving local political leaders more vulnerable if they helped coalition forces against insurgent groups. All of these actions surrendered the technical advantage enjoyed by U.S. forces to the political objective of lowering casualties in the short term.

The first battle of Fallujah (Operation Vigilant Resolve) in 2004 ended without a strategic plan. In this case, U.S. forces battled their way into Fallujah but ended up turning the city over to Iraqi security forces rather than taking control of the city and ensuring that insurgents did not return.[111] As should have been expected, Iraqi security forces could neither control the insurgents nor protect the population. Between April 2004 and November 2004, the U.S. military chose not to contest control of the city in the vain hope that the Iraqi security forces could somehow enact a sound counterinsurgency strategy, but the task was beyond them. The technical advantage of U.S. forces was surrendered for no gain.[112]

U.S. technical superiority was used to good effect in the second battle of Fallujah, or Operation Phantom Fury, in late 2004 and wrested control of the city back from the insurgents. However, Operation Phantom Fury was a large-scale conventional confrontation and *not* an example of good counterinsurgency practice. It was yet another conventional response to the lack of a sound counterinsurgency strategy prior to both battles of Fallujah.

Consider the words of Agence France-Presse photographer Patrick Baz, who was in both Fallujah and the 2004 battle of Ramadi shortly afterward: "At the time it was bad. No one even knew who the insurgents really were. In Ramadi we drove out of the fort and had to fight to the outpost. I didn't take pictures because I was passing ammunition to the soldiers. No one talked to the insurgents. But then I heard the colonel saying they would kill all insurgents. I thought is this guy serious? It is the whole city. That is when I knew it was time for me to leave."[113] When asked what he thought the difference in U.S. strategy was between Ramadi in 2004–5 and Diyala Province in 2008, he laughed and said, "Now they understand. They may not win. But, yes, now they see what they did wrong."[114]

V. Operationally Offensive, Tactically Defensive Strategy in Iraq

The operationally offensive, tactically defensive doctrine was used in Iraq to a lesser degree than it could have been. The concept was never enacted as a full-blown doctrine to be employed whenever appropriate, and it was not part of the force protection posture before the surge and General Petraeus's strategy to secure the population from the insurgents.[115]

While establishing large bases in hostile territory and posturing the forces in a defensive role could describe the force protection measures taken in Iraq before the Petraeus transition, taking these measures does not comply with a counterinsurgency doctrine chiefly because they do not have a counterinsurgency component. Furthermore, such a posture surrenders any initiative to the insurgents. It is the equivalent of burying one's head in the sand even if it is done in an operationally offensive, tactically defensive manner.

The early years in Iraq demonstrate that in counterinsurgency, much like technology, a core competency must come, first and foremost, with the application of an appropriate strategy. Only a sound counterinsurgency doctrine, with the appropriate technical asymmetry between the forces of the host nation and the insurgent forces, can provide the context in which operationally offensive, tactically defensive operations are viable.

Petraeus's surge was, in fact, an attempt to meld the technology, posture, and strategy into a close approximation of an operationally offensive, tactically defensive concept for counterinsurgency in Iraq. During and after the surge, the construction of small combat operations bases (COBs) was an attempt to move troops out into populated areas, where they could actually interact with and protect the population.[116] However, as noted in Diyala Province, the strategy still required incremental expansion of controlled territory to continue bringing more areas under the control of the government of Iraq. It is the incremental approach that separates the expansion of COBs in Diyala Province from the full potential of the operationally offensive, tactically defensive concept. The Fourth Brigade, Second Infantry Division, preferred a more measured approach in urban areas while accepting insurgent control of the countryside, with the inevitable cost of continued IEDs, suicide bombers, and car bombs trickling into cities, albeit at a greatly reduced rate.[117]

Small outposts, such as the COB DMC outside of Abu Khamis, arguably represented the maximum realistic expansion of the brigade's area of influence. The DMC was located on the eastern fringe of the southern prong of the Y, and the road leading to it had to be regularly cleared of IEDs.[118] If COBs were pushed out even farther into the rural areas, the number of dangerous roadways susceptible to IED emplacement would exponentially increase.[119] While Petraeus was on the right track, though possibly limited by technology, this situation illustrates another reason why both his and Colonel Lehr's troops were fundamentally tied to logistical support by road.

Using roads in Iraq was unavoidable, given the amount of matériel and number of vehicles a modern combat brigade requires. But an outpost would not have been necessary if that brigade had moved from company size to smaller formations, such as platoons or even squads.

The required movement of supplies then could have been substantially reduced as smaller units further integrated themselves into villages in more rural areas and largely lived off much of the local economy. A knee-jerk reaction may be that that expectation was not reasonable and that such a move would be too dangerous. Yet U.S. Army Special Forces Operational Detachments Alpha have been doing exactly that for almost fifty years.[120]

It is true that U.S. Army Special Forces teams have extensive training and have personality types that are more capable of integrating into other cultures and living off the land or with local communities. However, that should not preclude using conventional forces in a similar role to achieve wide coverage of rural areas, a tactic that would have benefited the Fourth Brigade, Second Infantry Division, in Diyala Province. Living in a fortified compound in a village, helping locals, and relying on technology to trade large areas of time for space are not the same work as training and leading up to battalion-size local militias in unconventional warfare, as expected of special forces teams. A combat medic, for example, does not need training in veterinary medicine and dentistry to help villagers. Such an individual could be fully capable of helping villagers between the occasional visits by special forces medics, who then would enhance the more limited skills of the conventional combat medics.

In Diyala Province, Colonel Lehr's focus on sound counterinsurgency technique, and his constant reiteration of their counterinsurgency role to his officers and noncommissioned officers (NCOS), produced obvious results.[121] However, his efforts failed to go as far as they could have because force protection and aversion to risk still were too commonplace. Implementing an operationally offensive, tactically defensive counterinsurgency strategy would have cost many more lives than the brigade lost. The increased fatalities would have had an inevitable political and financial cost although with a likely faster rate of progress in its counterinsurgency role.

Employing the operationally offensive, tactically defensive concept probably also would have exceeded the available air support in Iraq. It perhaps would not have been the case if it had been tried in a single province, like Diyala, but certainly if it had been tried countrywide. As previously noted, UAV support was already strained to the point that Shadow UAVs were flying double their design hours. Dispersing small units would have resulted in greater demand for these assets in Diyala Province, as well as for their control to be passed to lower echelons as specific situations required attention.

It should be noted from the Iraq example that using an operationally offensive, tactically defensive counterinsurgency strategy does not have to be an all-or-nothing affair. As every counterinsurgency is different, the counterinsurgency forces would require different approaches and flexibility, using different strategies in different phases. Colonel Lehr's incremental approach could initially have been well suited as the insurgents sought pitched fights (like the battle of Baquba) and did not move back to Mao's first stage and deny combat to U.S. forces (as they did after they lost control of the Y).[122] The strategy could then have shifted from the incremental approach to the operationally offensive, tactically defensive approach as the insurgents retreated and avoided combat, thus forcing them to fight again or lose their rural support. This level of adaptability was not evident in Diyala Province and may have been simply too much to demand of the U.S. military at that time given the politically sensitive issue of casualties.

VI. Iraq Holistically

Hindsight makes analysis easier and perhaps, at times, uncharitable. It is easy to criticize the inadequate planning during the prelude to the war. Criticism may be justified and rightfully attributing strategic deficiencies to policy makers, but it may not be equally valid to criticize the U.S. military and its failure to come quickly to grips with the insurgency it was facing. It may also be inaccurate to classify the 2004–7 period as wasted or without strategy. As David Kilcullen notes, "The Iraqi Arabs do not live for fighting. They are often cruel but do not like to actually fight."[123] The period prior to the surge offered the Iraqi militias plenty of opportunities to engage the U.S. military in direct fire engagements, and they lost every engagement. This period may have served as a valuable lesson on the lethality of confronting U.S. forces and leads to some interesting questions. Would the Awakening Councils/CLCs have cooperated with counterinsurgency forces after they had rid themselves of al-Qaeda if they did not know that continued direct combat with U.S. forces was so lethal? Would they have simply gone back to direct engagement with U.S.

forces if they did not fear them? Indications are that the British forces in Basra never achieved a deterrent effect and were subject to attack almost until the time they left.[124] Would the same have been true for U.S. forces if the violence and technology used in places such as Fallujah, where overwhelming firepower killed more than thirteen hundred insurgents, had not been demonstrated?[125]

It should never be argued that a lack of strategy is preferable to a bad strategy. But it is possible that the context of the war in 2006, and the seriousness with which insurgents regarded U.S. firepower, was a direct result of strategic neglect and the insurgency's growth to a point where it tried to engage U.S. forces. If this observation is true, then the opportunity to implement a partial operationally offensive, tactically defensive counterinsurgency strategy may have been missed, as a deterrent period may have provided the background for it to be more ably implemented.

Constraints do not seem to have affected the U.S. intervention in Iraq on a strategic level as restraint, in this case, helped further counterinsurgency strategy; however, constraints certainly had a negative effect on a tactical level. American troops often had to detonate IEDs by dropping grenades next to them and running away even though far safer alternatives were available.[126] One alternative, using electromagnetic pulse–generating platforms to detonate IEDs remotely, was banned out of concern that they might set off any IED in the area and civilians could be hurt.[127] Troops in Diyala Province in 2008 often complained that they could not return fire or call in fire before receiving authorization.[128] Every officer interviewed in Iraq for this book unequivocally stated that they believed the constraints cost the lives of soldiers.[129] This issue could have become much more problematic if Iraq again had slid back into sectarian strife and the soldiers had felt as though they were getting killed for a "lawyerly" reason.

Finally in the U.S. intervention in Iraq, the counterinsurgency struggle there will continue for quite some time into the future. The technology used allowed the United States to engage in a counterinsurgency struggle with far fewer troops than conventional wisdom deemed possible. The application of sound counterinsurgency strategy, when combined with technical advantages, allowed the U.S. military to achieve some modicum of success by the autumn of 2009.[130] But the reality is that Iraq remains fragile and the intervention cannot be considered a success unless and until the Shia and Sunni political blocks can manage to work together. The U.S. military or its technology or its counterinsurgency practice, however sound, will not decide that outcome.

3 Limits of Politically Correct Doctrine

The Melians: You may be sure that we are as well aware as you of the difficulty of contending against your power and fortune, unless the terms be equal. But we trust that the gods may grant us fortune as good as yours, since we are just men fighting against unjust.

. . . The Melians surrendered at discretion to the Athenians, who put to death all the grown men whom they took, and sold the women and children as slaves.

THUCYDIDES

I. Assumption in Counterinsurgency

Using the operationally offensive, tactically defensive concept in pursuit of counterinsurgency objectives relies on the assumptions that an insurgency will employ Mao's concept of revolutionary warfare and, more important, that the population of the host nation should be receptive to influence by counterinsurgency forces. These assumptions seem to be widely held in the historical literature on counterinsurgency, such as David Galula's *Counterinsurgency Warfare: Theory and Practice* and Gen. David Petraeus and James Amos's field manual *FM 3-24*.[1] The field manual specifically surmises that in most insurgencies about 15 percent of the population will be in favor of the insurgents, 15 percent will support the host nation forces, and 70 percent will be a neutral element.[2] As this chapter explores, such assumptions are extremely dangerous and present profound problems if the assumptions are correct and potentially catastrophic problems if they are not.

As *FM 3-24* points out, the counterinsurgency force must eliminate hard-core fanatics and insurgents primarily because the possibility that they can be coopted into supporting the counterinsurgency objectives is low. This effort, of course, means that such forces must be either eliminated or separated from the population through incarceration.[3] Ostensibly such fanatics would be classified in the first 15 percent of the population who would be assumed to favor the insurgency. It is quite likely that not all of the population who supports the insurgency could be classified as fanatical to the point that they must be eliminated or incarcerated. Moreover the basic assumption is not going to be relevant in every theater of warfare or every counterinsurgency conflict. Undoubtedly at times the percentage of the population supporting the insurgents, or supporting the host nation, might well exceed the 15 percent mark. To further complicate the issue, the percentage of non-repentant fanatical elements in the population who support the insurgency could vary widely depending on the society and state of conflict in which counterinsurgency operations are taking place.

To assume that 15 percent of the population favors the insurgency and that perhaps only 5 percent represents a fanatical element that must be eliminated or incarcerated means that the sheer number of individuals involved could cause a strategic problem. Consider the case of Iran in 2002. With a population of more than 60 million people, the 5 percent who would have to be incarcerated or eliminated would amount to 3 million individuals.[4] Any attempt through combat operations to eliminate 3 million individuals, willing to die for their cause, would result in inevitable casualties for the counterinsurgency forces. In very basic mathematical terms, a 10:1 exchange ratio between host nation forces and enemy insurgents would result in 300,000 counterinsurgency deaths. Such a death rate for U.S. forces engaged in any conflict is simply beyond the political capability of sustainment for a liberal society not directly threatened with its own survival. Even at a 100:1 exchange rate, the resulting 30,000 casualties would represent seven times the casualties sustained in seven years of fighting in Iraq.[5] At a 1,000:1 exchange rate, the 3,000 casualties could possibly be politically acceptable in the United States. However, this possibility assumes two important points, both of which may be flawed: there would be no more than 5 percent of the total Iranian population who would provide fanatical support to an insurgency, and the United States would be willing to achieve a 1,000:1 or better exchange rate against insurgents.

There is a direct corollary between the percentage of the population who represents non-repentant fanatical elements that must be eliminated and the exchange rates that the U.S. military must achieve in order to make a counterinsurgency effort viable in a political context that would be acceptable to both Congress and American citizens. Simply adjusting numbers starts to drive home just how skewed the exchange rates must be. Consider for a moment Indonesia, which is unlikely to be a target for U.S. military intervention but does represent a Muslim state that in the advent of profound social upheaval could pose a threat through its support of al-Qaeda. Indonesia had a population of 212 million people in 2003.[6] If 5 percent of the Indonesian population had to be eliminated or incarcerated, serious questions must arise about the viability and political cost of treating 10.5 million human beings in such a fashion. And what happens if 10 percent, or larger numbers, of people are involved? Such a conflict starts to illustrate the true difficulty in a counterinsurgency effort of this size.

The difficulties in these illustrations, however, should not be taken to mean that counterinsurgency is impossible in such areas or against such large populations. There is the possibility that the United States will feel compelled to retaliate against a group of insurgents or terrorists embedded in a larger population. Afghanistan is a case in point. The United States, while having no inherent strategic need to confront the Taliban, could not allow the Taliban to harbor al-Qaeda in Afghanistan after the attacks on September 11, 2001. In a situation where the United States feels that it must act, it must be careful in how it proceeds and tailor the counterinsurgency strategy appropriately to fit the political and social milieus in which it is fighting. In the case of the Taliban's harboring al-Qaeda in Afghanistan, the U.S. strategy required that the Taliban be, at least temporarily, considered an enemy force.

The most daunting military challenge facing the United States is its counterinsurgency effort against the Taliban in Afghanistan. Unlike al-Qaeda, the Taliban themselves were not perpetrators of aggression against the United States. The recognition that the Taliban, despite cultural influences and the historical ties from fighting the Soviet Union with U.S. aid, were simply harboring al-Qaeda means that the United States has a very hard time making the case to the Pashtun people that the Taliban and al-Qaeda are their enemies. The argument that the Pashtun people should strongly support both a central government, which is not historically important to them, and a mainly Christian army, such as the U.S. military, against their own countrymen is unlikely to be given much merit.[7] Since the Taliban is primarily Pashtun, and often supported by the Pashtun population, the United States should not generally count on Pashtun assistance in the hunt for al-Qaeda and the effort to sustain a central government.

With the Taliban enjoying solid backing in the Pashtun tribal areas of both Afghanistan and Pakistan, it is very unlikely, though predicting the future is always hazardous, that any country will be able to convince the Pashtuns to back outside and alien political objectives. In such a situation, a counterinsurgency force is presented with three options. The first option is simply to walk away from the conflict and accept failure in achieving the political objectives. Clearly this option is going to be dependent on U.S. domestic political influences, which will be different for every given conflict. For example, the 1993 Operation Restore Hope in Mogadishu, Somalia, enjoyed far less political and popular support than the 2001 invasion of Afghanistan did.[8] Given the incentives for making sure that Afghanistan does not again become an al-Qaeda stronghold, the U.S. government will unlikely accept walking away from Afghanistan as a realistic option; however, the historical example of the Soviet experience in Afghanistan shows that a counterinsurgency effort that loses public or political support can be terminated.[9] The second option is to adjust the narrative that defines the conflict in order to adjust the political objectives. Any insurgency should be a dynamic learning situation and, as such, should present opportunities to adapt to new challenges. Likewise, the perception of a counterinsurgency effort can change with regard to the viability of the original political objectives, which, later in the conflict, may then be transformed and no longer be applicable. In such instances, if accepting defeat and the withdrawal of forces is not an option, it may become possible to change the narrative of the counterinsurgency effort to adjust the political objectives to more realistic possibilities. Finally the third option is to apply force against the civilian populations who support the insurgency in order to threaten the insurgency's lifeline. Such an option would entail substantial political costs for the United States internationally and would also challenge U.S. law. Unrestricted operations of this kind are not possible for the U.S. military.

II. Withdrawal

If the United States finds itself in a counterinsurgency effort in which the majority large portion of the concerned population actively supports the insurgency, any hope of victory, or of obtaining the political objectives, becomes negligible. It is extremely hard for the American psyche to understand how other societies and countries would not want to emulate the United States, given its preponderance of military and economic power and its standard of living.[10] Most Americans believe that liberal democracy is superior to all other systems of government and that this fact should be universally self-evident. Such hubris, and often ignorance of other cultures, prevents many Americans from being able to understand and appreciate that other cultures and societies are satisfied within their own context. It follows quite naturally that without an appreciation of other cultures, strategy formulated by Americans will lead to assumptions that develop into policy objectives that are unrealistic in a counterinsurgency environment.

The United States responded to the terrorist attacks of September 11, 2001, by demanding that the Taliban should hand over members of al-Qaeda to U.S. custody. The Taliban, given their cultural attitudes toward giving sanctuary to fellow Muslims, refused the request. This denial set the stage for the U.S. invasion of Afghanistan.[11] Of primary importance here is the notion that the United States was interested in killing or capturing members of al-Qaeda, not the Taliban. In essence by asking the Taliban to hand over al-Qaeda, the United States was telling the Taliban that it was not interested in entering into armed conflict with them and would do so only as a last resort based on Taliban noncooperation. The Taliban's refusal may have been a calculated strategic decision on their part or a religious decision not to cooperate with the infidel. The end result was that the United States gave support to the Northern Alliance of Afghanistan.[12] This forced partnership has led to a long counterinsurgency campaign in which the Pashtun people will determine whether the United States will accomplish its goals.

As the conflict in Afghanistan has progressed, the narrative of the conflict has changed. While it was originally about killing or capturing members of al-Qaeda in response to 9/11, it has since developed into a counterinsurgency struggle against the Taliban. The strategy also focuses on a concurrent effort to establish a secure state in Afghanistan with a strong central government as a bulwark against the recurrence of a terrorist sanctuary for non-state actors.[13] It makes sense at the strategic level to ensure that al-Qaeda will not again use a lawless area of Afghanistan to train, equip, and plan for terrorist attacks in pursuit of its global jihad. But such an objective implies a strong central state capable of policing its borders and exercising control within them. Afghanistan, today and historically, has never had a strong central government and has no institutional inclination toward developing such a government.[14] Concentrating power in the hands of a central government entails removing power from regional actors, such as warlords and tribal leaders, who will counter that political objective with hostile violence against both host nation and counterinsurgency forces. Ergo, the moment the conflict in Afghanistan changed from a raid to kill or capture al-Qaeda members to a counterinsurgency effort to suppress the Taliban and build a strong central government in Kabul, the entire narrative and implications of the war significantly changed.

There is a drastic difference between a raiding strategy meant to kill or capture and a long-term counterinsurgency effort. The U.S. military's technical superiority makes a raid that is small in size, or even large in scale, fairly easy to accomplish. However, as noted in previously, the technology required to use an operationally offensive, tactically defensive concept effectively with very small units to cover the maximum possible number of population centers is not yet mature. From a technical aspect, the moment the narrative changed in Afghanistan, the U.S. military had to shift from engaging in a raid, in which it had every tool it needed to succeed, to facing a long-term counterinsurgency struggle. In this context it had some of the technical tools to be successful but with very little long-term prospect for success because of previously mentioned legal and cultural restrictions. The drastic change in strategy and in the narrative equates to a drastic change in the likelihood of success in obtaining the political objectives.

Ultimately the United States must recognize that a long-running, grinding counterinsurgency campaign against a population who supports insurgents is doomed to failure with the legal restrictions currently in place that limit the use of U.S. military power. While it is true that the U.S. military has the technical tools to withstand any amount of direct combat pressure and has the ability to inflict massive casualties on the insurgents who choose to confront the U.S. military directly, the simple fact is that the U.S. military cannot stay deployed in Afghanistan forever. Likewise, while technically inferior and likely to die when engaged in direct combat operations against a technically superior U.S. military, the insurgent can simply play for time while using asymmetrical methods of warfare. To illustrate the problem, suppose for a moment that the U.S. military had all the technical tools needed to implement a wide-ranging operationally offensive, tactically defensive concept to secure even the smallest Pashtun village against insurgents. The result would be the same if the people could not reconcile themselves to being loyal to the host nation or to working with the counterinsurgency forces. In such a situation, the U.S. military's criterion for victory or for obtaining the political objectives is to have the willpower to outlast its opponent or to impose its political will on the population. This strategy is simply not feasible, however, in the context of the U.S. political milieu. While the point may be argued, it should be self-evident that a farmer fighting for his land, his family, and his identity will keep fighting and stay motivated longer than would a fickle American population thousands of miles away with other concerns, especially if constraints ensure that the farmer is unlikely to come to harm.

When the United States is confronted with such a stubborn counter-insurgency situation, it can cogently be argued that withdrawal from the conflict is in its best interests. That action would clearly call for sacrificing the original political objectives driving the conflict and accepting the failure to achieve those political objectives. Any such action will have profound ramifications in the American cultural experience and for the careers of the individuals involved in making such a decision. But at a strategic level, such a decision may well be the best option given the political and financial costs associated with changing the conflict's narrative or in unrestricted operations.

III. Changing the Narrative

In November 2009 Secretary of State Hillary Clinton stressed that the U.S. objective in Afghanistan was to kill or capture the members of al-Qaeda who were responsible for the September 11 attacks in the United States.[15] While the administration of President Barack Obama had revised its strategy for Afghanistan but had not yet solidified it into concrete gains, Clinton's articulation could be classified as an effort to change the narrative in the Afghanistan conflict. While it is impossible to be certain, the changing narrative was probably in recognition that a long-term counterinsurgency struggle against the Taliban was unlikely to produce political results that were acceptable to the Obama administration, such as the withdrawal of U.S. forces by 2011. However, it is quite odd that mission creep should be allowed to happen after almost eight years of conflict.

Fundamentally, the voting population of the United States determines the viability of a long-term counterinsurgency struggle conducted by U.S. military forces. Clearly the forces are under the direct command of the president of the United States, who most likely closely consults with the military and members of Congress, but the voting public ultimately has the final say on the existence of continued military operations. In the case of the Vietnam War, the withdrawal of U.S. forces can be directly linked to the disillusionment of the American populace with the cause in question. It is likely that the war in Iraq would have been terminated and the withdrawal of U.S. troops begun earlier if marked improvement had not become evident in 2007 and 2008. Again the withdrawal of U.S. troops would have been the result of plummeting popular support for the war rather than the reduced political and financial capability of the United States to continue operations in Iraq. The war in Afghanistan followed a similar path: progress had to be perceived and perception translated to political reality to keep a long-term commitment to obtaining political objectives in that counterinsurgency environment.

That the American populace had to see progress in Afghanistan put the Obama administration in an extremely difficult situation regarding determining the best strategy for moving forward with the conflict. Owing to the war's legitimacy, given its catalyst of 9/11, a withdrawal of U.S. forces could lead to a catastrophic political defeat for the Obama administration.

On the one hand, the American people may come to view the Democrats as weak and as having lost the war in Afghanistan. On the other hand, as noted, it may be strategically sound to withdraw given the alternatives and the long-term prospects for obtaining the political objectives without resorting to unrestricted operations. Any attempts to continue the counterinsurgency effort could result in blame for the administration.

Likewise, increasing the forces in Afghanistan, as requested by Gen. Stanley McChrystal in the autumn of 2009, resulted in the reinforcement of failure at the strategic level in the Afghan conflict. The population of southern and southeastern Afghanistan was not co-opted through counterinsurgency strategy by applying an operationally offensive, tactically defensive concept in order to leverage technology. The reinforcement of conventional troops into a counterinsurgency environment has led to increased political and financial costs, and withdrawal has become the more acceptable option to the American people. But a refusal in 2009 to reinforce the troops, however strategically sound that may have been, would have set the Obama administration at odds with the U.S. military and incurred additional political costs.

Given that neither withdrawal nor escalation are suitable options in a counterinsurgency environment such as Afghanistan, when the populace cannot be co-opted, the only option left to the Obama administration is to change the narrative of the conflict and thereby adjust the political objectives to make them appear to be attainable. In the case of Afghanistan, the administration attempted to adjust the narrative, to divide the different factions that constitute the Taliban, and to isolate al-Qaeda. By emphasizing that the United States intended neither to stay in Afghanistan forever nor to rebuild the state of Afghanistan to resemble liberal Western democracies, Secretary Clinton had attempted to shift the narrative from a counterinsurgency effort against the Taliban back to the original political objective set after September 11, 2001—to kill or capture members of al-Qaeda.

By adjusting the narrative to split the more moderate factions of the Taliban away from the main insurgent effort, as well as to isolate al-Qaeda, the administration hoped to create strategic movement and generate increased support from the American people to salvage the original political objectives of the conflict. It remains to be seen whether the Taliban are secure enough in their current strategic trajectory to close ranks with their al-Qaeda allies and maintain cohesion in the face of an adjusting narrative from the United States. Two further interesting possibilities came from the adjustment of the Obama administration's narrative of Afghanistan in late 2009. Emphasizing the need to kill or capture al-Qaeda rather than engage in nation building indicated a preference for a raiding strategy rather than waging a counterinsurgency struggle. It also assumed that the U.S. military had visited enough violence and fear on the Afghan populations in southern and southeastern Afghanistan to make them amenable to a peaceful arrangement with the United States by which they would agree to stop supporting al-Qaeda. Unfortunately, it is not clear that either of these two points is accurate.

It was quite interesting to see an administration, while not explicitly supporting a raiding strategy, adjust the narrative of the conflict and lead to the conclusion that a raiding strategy would be preferable to the more liberal notions of spreading democracy and enlightenment through the promotion of good governance and a social contract. Changing the narrative in Afghanistan to focus on al-Qaeda rather than the Taliban may have only been a ploy in order to have the Taliban voluntarily separate from al-Qaeda rather than an indication that the U.S. military was capable and willing to mount raids and use violence to target al-Qaeda. However, it was logical to assume that the straightforward statement by a U.S. government official as administratively important as the secretary of state was not a bluff. Unfortunately liberal politicians, not trained in strategy, likely did not comprehend the level of violence that raiding strategies employ in order to be successful.[16]

As a further note, any attempt to change the narrative of a conflict in order to divide an enemy faction or disrupt cohesion must be predicated on violence and fear used against one of the factions. Without the application of violence and fear, faction members have absolutely no reason to divide their loyalties or become susceptible to division and hence diminution of their power and survivability. While battles in southern and southeastern Afghanistan have cost the Taliban and local people high casualties, thus far, within the Afghan social context, such casualties have not constituted a motivation to continue fighting as a cohesive force or for isolating al-Qaeda and accepting and participating in a central Afghan government.

IV. Unrestricted Operations

Counterinsurgency operations are not new to the twentieth century. Guerrilla warfare has always been a part of warfare and is simply an asymmetrical cost-effective way to confront a larger and stronger opponent.[17] Yet counterinsurgency operations and doctrine received new life and rose to prominence since the 1950s for a very peculiar reason. Simply put, the liberal way of warfare, which is how Western liberal democracies conduct themselves in warfare, is ultimately what has allowed ideas such as Mao's revolutionary warfare or insurgencies to become widespread.

The very nature or essence of warfare—that is, the violence and fear perpetrated against others to attain a political objective through the imposition of political will—seems to be beyond the grasp of most people in Western societies.[18] It is important to note that morality does not play a part in the nature of warfare itself; rather, morals are applied in different ways and through different mechanisms by different societies to regulate or constrain the activity of individuals engaged in warfare. This, however, is not the same as recognizing the actual nature of warfare. Because warfare in a pure form is predicated on violence and fear, it is an inherently amoral or immoral activity; by its nature it cannot be moral in any absolute sense. Of course, war can be justified on moral grounds; in fact, such rationalization is the norm, especially for the United States, as justifications for its actions are required for domestic support.[19] It should also be noted that many academics without combat experience or firsthand observations of wound ballistics would strongly take issue with the assertion that war cannot be moral in any absolute terms. In a Western context it is extremely important that every war be given a moral foundation to justify it to the people who hold liberal democratic ideals to be proper.[20]

It is difficult to formulate appropriate strategies for some counterinsurgency situations. Morality complicates the issue further, as what is considered moral is constantly changing but generally becomes less flexible at all levels of warfare. For example, food control and rationing practices by British forces in Malaya during its communist insurgency of

1948–60 would today be considered, at a minimum, immoral and possibly a crime under international law. Through the use of global communications such as the Internet, some practices—such as the U.S. strategic hamlets initiative in Vietnam in the 1960s, or the internment of Vietnamese in concentration camps—now would be portrayed as ethnic cleansing. While that classification would be very unjust, it does not alter the fact that such programs can be portrayed with a moral spin through available global communications to turn almost any issue into propaganda.

Given the historical success of dealing with counterinsurgency by imposing collective punishment on the population who supports the insurgents, it could be employed again in the future. Collective punishment through the severe repression of a population in order to suppress an insurgency can and does, in fact, work. Examples include the Russian involvement in Chechnya, the Sri Lankan campaign against the Tamil Tigers, and the Sudanese use of militias in Darfur. All visit violence and fear upon the population to ensure the people understand that they will pay a cost for supporting insurgents, one that is far higher than what the insurgents can impose for cooperating with the host nation. Such campaigns often elicit moral outrage, strongly worded objections, and much condemnation from Western governments due to the moralistic lens that these liberal democracies use when judging a conflict.[21] The moral outrage often voiced by Western powers, however, does not alter the fact that such methods, while brutal and violent, are effective strategies for dealing with insurgencies.

It should further be noted that current moral attitudes do not support strategies that call for collectively punishing the civilian population in order to dissuade insurgency and that this book in no way advocates reexamining the laws or norms that prohibit such behavior.[22] However, in all likelihood, if the United States were engaged in a sufficiently threatening war, the rules and laws governing the behavior of the U.S. military, with the moralistic overtones that are present today, would very likely be adapted to properly fit the context or narrative of such a conflict. Further if sufficient violence such as a nuclear or biological attack were visited upon the United States on a repeated basis, the country's obligation to respond in a moral way would become far less important than it is today; the population would demand retribution from their political leaders and the military. Any such demands, however, should be carefully considered as it is imperative that the United States behave in a way that conforms to international laws and norms.

It is conceivable that the United States could wage counterinsurgency operations in several parts of the world where the population could not be won over to supporting the host nation. In theoretical U.S. military interventions in southern Lebanon, Iran, Yemen, Saudi Arabia, Pakistan, Somalia, and Afghanistan, the United States would have a very hard time winning hearts and minds. In these instances, conducting a counterinsurgency strategy following Mao's three stages of revolutionary warfare to defeat the insurgents would be almost impossible.[23] After entering into a conflict and intervening with whatever forces are appropriate, the United States would be faced with a long-term insurgency that would most likely be supported by the local population.

The U.S. military does not always have to fight an insurgency with a long-term counterinsurgency plan and doctrine such as the use of the operationally offensive, tactically defensive concept. The United States is very capable of defining conflicts outside geographical borders with the emphasis on the state structure of the given territory. Rather than engaging in long-term military operations, the United States may in the future choose to use its technical advantage and reach to deal with potential adversaries with a raiding strategy rather than with a counterinsurgency strategy.

Historically using raiding strategies for punitive effect to build deterrence against an insurgent movement or groups using guerrilla tactics was a viable option and a part of small war theory as outlined by Col. C. E. Callwell for dealing with recalcitrant populations in Afghanistan.[24] Further the raiding strategy is still a viable option, as seen with the early 2009 incursion of the Israel Defense Forces into the Gaza Strip to reduce or eliminate indirect fire from Palestinian militants. The main difference between a historical use and a modern use of a raiding strategy can be seen in the aggregate political cost associated with using violence with or near a civilian population. In the late nineteenth century and into the early twentieth century, raids may have elicited some sense of moral indignation in those populations throughout the world who had access to information. However, technology has fundamentally changed the viability of such strategies with the worldwide dissemination of information and news in real time. Ironically as technology has allowed an increased military capability, especially for the U.S. military and its technically oriented posture, technology has also allowed for a revolution in communications for disseminating propaganda. For example, a raid into enemy territory resulting in fifteen hundred dead enemy combatants and civilians, historically, would have been nothing more than a footnote in a book or perhaps a passing article in print in the late nineteenth century. By 2009 in the politically charged environment of the Gaza Strip, for example, the widespread availability of picture-capturing cell phones and camcorders and of the Internet to broadcast the real-time images has resulted in large-scale awareness and the constant circulation of emotionally charged images with high propaganda value for Palestinian militants.[25]

If the United States finds itself engaged in a counterinsurgency effort in which its populace is unlikely to be convinced that supporting the host nation's government is in its best interests, and withdrawal or changing the narrative of the conflict is not possible, then the United States will have to be very aware of and grapple coherently with the resulting negative propaganda that will ensue. Ultimately if the United States were to engage in punitive raiding measures, or unrestricted operations, then it would possibly be forced to surrender the moral high ground that it desperately seeks in any confrontation.[26] This yielding of the moral high ground, of course, may not be as problematic as it sounds given that the United States does not enjoy the moral high ground in the eyes of most individuals living in the Muslim world, where much of the intervention is likely to take place.[27] The real challenge for the United States will be how to cope domestically with the loss of the moral high ground and the implications of those losses vis-à-vis political relations with other nations in the world, especially western Europe. Even domestically such a shift away from moralistic warfare and rules of engagement is likely not possible or desirable.

V. Afghanistan

The Soviet invasion of Afghanistan in 1979 resulted in an unexpected long-term counterinsurgency struggle that ended in failure for the Soviets.[28] The common perception of this conflict, especially in the Arab world, is that the holy warriors, or mujahideen, with the help of Allah, managed valorously to throw back the atheistic Soviet war machine, resulting in the Soviet Union's dissolution.[29] Naturally the mujahideen who fought against the Soviets in Afghanistan picture themselves, along with their faith, as being the principal reason for the Soviet military's defeat. And while it is true that the Soviets would not have been defeated without the mujahideen's willingness to sacrifice their lives, personal valor is simply not the whole story. The common belief that technology is not a decisive implement in counterinsurgency strategy proved false in Afghanistan as it did become the deciding factor in the outcome of the conflict.[30] It should be noted that the Soviets' use of extreme oppression and unrestricted operations against those willing to support the mujahideen was a successful tactic until technology changed the conflict.[31]

Initially the Soviet Union had considerable trouble dealing with the mujahideen, who preyed upon the Soviets' logistics lines while using guerrilla tactics to inflict casualties.[32] The Soviet military's primary mode of operation was to use motorized rifle units moving with substantial artillery support and attempt to close with and kill or capture insurgent fighters.[33] It quickly became apparent that this strategy left the Soviet military extremely vulnerable and quite often unable to take effective control of substantial areas of territory. The Soviets, however, had considerable technical superiority over the insurgent forces, especially the ability to use close air support in both tactical situations and for punitive measures against population centers that supported the insurgents. Relying upon ground forces alone, moving through constricted valleys and passes, was simply not an effective method of fighting in the mountainous country of Afghanistan. The Soviet Union developed new tactics that employed air mobility to quickly move forces in theater.[34] This doctrine evolved into hammer and anvil operations in which an airmobile force would be moved into a valley while a ground force would maneuver to squeeze the civilians and the mujahideen between them. It became clear to the Soviets that doctrinal change to incorporate air mobility would serve them better in mountainous terrain than using motorized rifle divisions.[35]

With air mobility and close air support, the Soviet forces were able to effectively dislodge the mujahideen, who would disengage from them and melt away into the hills and side valleys. The Soviet military achieved tactical success on the battlefield but strategic success would continue to elude them. The battles for the Panjshir Valley were typical of this type of operation. In seven successive battles in the Panjshir Valley, the Soviet forces would push their way up the valley, engaging and defeating the insurgent forces, who would then slip into the side valleys and out of contact with the Soviets.[36] The Soviet forces would not be able to sustain themselves in the valley indefinitely and would have to withdraw, allowing the insurgent forces to trickle back from the side valleys and once again harass the Soviet supply line from the Salang Highway.[37] Quite similar to the U.S. experience in Vietnam, the Soviets would find themselves battling for a given area, using their technical and firepower superiority to be successful at the tactical level, and then simply withdrawing after their tactical success without taking a strategic advantage.

It became clear to the Soviet Union that defeating the insurgents would require ending their popular civilian support. An outstanding analogy is Mao's often-quoted description of the insurgents as fish and the population as the water in which the fish swim. However, lakes can be drained. The Soviet military forces instituted a campaign of terror against the civilian populations of villages that supported insurgent efforts.[38] While their actions were morally abhorrent to the West and many, especially Europeans, claimed that they violated international law, the Soviets may very wisely have come to the conclusion that the Afghan population would be unlikely to accept, or support, an atheistic government. Such a political system and way of life would be historically anathema to everything that the Afghans' own personal identities held dear. The Soviet Union then had the option of withdrawing its forces, changing the narrative (which would have been difficult), or engaging in punitive operations to annihilate or subjugate the populace supporting the insurgents. Many writers of counterinsurgency theory and practice stipulate that the latter actions actually enrage the populace and result in the inevitable defeat of the counterinsurgency forces as insurgents have an almost limitless manpower supply. As the Soviets clearly understood, this view is not, in fact, always accurate.

It is true that brutal repression of a population will engender negative feelings and help an insurgency to recruit like-minded individuals.

Typically and especially in a culture such as Afghanistan's where revenge (*badal*) is seen almost as a duty, the killing of a civilian is likely to generate more enemies.[39] The conventional wisdom that repression will ensure defeat, however, only holds true as long as counterinsurgency forces are unwilling to kill every last member of the population who supports the insurgency. The almost tacit understanding is that indiscriminate use of force is countereffective only because the United States and Western forces are unwilling to kill an entire population or enough of a population that its remaining members understand that unless they stop supporting the insurgency they too will die. In recent conflicts, it has been quite evident that the U.S. military, while being extremely capable of doing so, has not been willing to annihilate a population, nor is it likely to be willing to do so without a profound shift in legalistic and moral attitudes in the United States. This observation does not change the fact that in a historical context the use of unrestricted operations, or a raiding strategy to exact punitive measures on the population, has been an effective way to deal with insurgencies.

While the common perception is that the Soviet Union entered into a quagmire in Afghanistan, it is not widely understood that the Soviet Union was, in fact, very much in control and that both its own military and the Ronald Reagan administration expected it to be successful in Afghanistan.[40] By 1984 the head of the Inter-Services Intelligence (ISI) of Pakistan had similar misgivings about the likelihood of success in effectively defeating the Soviet military.[41] In fact Brig. Mohammad Yousaf, the head of the ISI's effort to arm and equip the mujahideen of Afghanistan, was convinced that the insurgency was doomed to failure unless more modern weaponry was made available to the mujahideen to counter the technical dominance of the Soviet Union. By late 1984 the Soviet military, using ruthlessly repressive measures, had driven much of the insurgents' civilian support structure across the Pakistani border, with 3.5 million refugees seeking shelter in Pakistan and 1.5 million refugees seeking shelter in Iran.[42] The people of Afghanistan were being either repressed to the point of compliance or driven out of the country. This period in the Soviet-Afghan conflict is often ignored as it neither reflects the ideal prowess of the Arab or Muslim mujahideen nor portrays the half-hearted U.S. support to the mujahideen particularly favorably.

It became quite clear by 1984 that to have any chance of success, the mujahideen required an ability to negate the threats of Soviet close air support and Soviet helicopter gunships, such as the Hind armed-assault helicopter, which were making punitive raids on Afghan villages without repercussions.[43] Until that point, the mujahideen had been supplied with SA-7 and Blowpipe shoulder-launched surface-to-air missiles to counter the Soviet air threats.[44] Both of these weapons proved to be completely inadequate for the job. Specifically while the SA-7 was available in fairly large numbers, it had the limitation of needing to be generally above and always behind the target to be effective. Such targeting opportunities are exceedingly rare.[45]

At that time the United States had in its military inventories the Stinger surface-to-air missile, which was truly a state-of-the-art weapon; however, the United States had not been willing to provide the insurgents with such technically sophisticated weapons.[46] Giving this weapon system to the Afghan mujahideen would almost guarantee the technology's dissemination to Iran, the Soviet Union, Pakistan, China, and any other third party willing to purchase the technology. The United States ultimately decided to provide the mujahideen in Afghanistan, through the equipment pipeline of Pakistan, with Stinger missiles to counter the Soviet air threat.

The technical ability of the Stinger lies in its advanced seeker head, which after acquiring the target, guides the missile from launch to impact. While the SA-7 is generally only capable of locking on to the aircraft's engine exhaust from behind, where the heat source is extremely visible, the Stinger has the ability to detect an aircraft's heat source head-on.[47] In other words, the Stinger surface-to-air missile is capable of locking onto a target regardless of the aircraft's direction of travel or the angle of that direction of travel relative to the missile's shooter. Under ideal conditions, therefore, any aircraft flying at less than ten thousand to twelve thousand feet above ground level is susceptible to being shot down by a Stinger missile regardless of the aircraft's altitude. The follow-on result is that aircraft flying at such heights, without modern precision-guided munitions, are almost wholly incapable of providing effective close air support.

The Stinger missile removed the viability of close air support as a tactical advantage for Soviet military forces in Afghanistan, thus changing the military strategic context in which the conflict was being fought. The first result was that Soviet aircraft, then unable to operate at low altitudes, could not effectively carry out punitive raids against Afghan population centers that were known, or suspected, to support insurgents.[48] Therefore, the mujahideen's supply bases, which in reality consisted of villages in remote areas of Afghanistan, became sanctuaries that could only be threatened by the use of ground forces, with the attendant risks of ambush and increased vulnerability of their logistics trains. The second result, and more important for the overall strategic context of the conflict, was that combat between Soviet military forces and the mujahideen took place as direct fire ground engagements or indirect fire to harass Soviet base areas. In this context the mujahideen were still at a tactical disadvantage to superior Soviet military firepower, but as was the case for the U.S. military in Vietnam, the overall effect was to transform the conflict into an attrition style of warfare. Given an almost endless supply of volunteer insurgent fighters, albeit still at a severe technical and tactical disadvantage, the mujahideen were capable of inflicting a constant stream of Soviet casualties.

Unlike the U.S. military today, the Soviet military in Afghanistan did not have adequate body armor to make small arms fire a minimal concern, nor did the Soviets enjoy widespread night vision capability to adequately exploit the possibility of nighttime operations.[49] The end result was an inevitable trickle of casualties from almost any engagement combined with the relinquishment of tactical initiative and tempo to the mujahideen after the sun set. Afghani, Pakistani, and foreign mujahideen volunteers from across the Arab world, while being supplied by Western, Saudi, and Pakistani sources, could conceivably continue to fight an almost inexhaustible battle from safe havens in Pakistan and Afghanistan.[50] Without Soviet resolve to eliminate the safe havens in Pakistan and, therefore, local support for the mujahideen, the Soviet military lacked long-term options for successfully prosecuting the conflict to victory, and it inevitably withdrew from Afghanistan.

From this summary, it logically follows that the Soviet defeat in Afghanistan was not, in fact, the result of poor counterinsurgency strategy or of heavy-handed tactics and brutal repression of the Afghan population. The Soviet Union faced a situation where it was incapable of sufficiently changing the narrative of the conflict so as to co-opt the Afghan people into a government system that was alien in its structure and that emphasized central control. Not only was the government structure foreign, but also it was based on an ideology that was utter anathema to the popular culture of most Afghans. In choosing a counterinsurgency strategy that involved the brutal oppression of the population, who supported the insurgents, the Soviets were simply opting not to withdraw and thereby following the only other available option. While using unrestricted operations against a populace who would never support them was, in fact, a sound tactic, it was not an appropriate one for the U.S. military when it entered Afghanistan.[51]

The Soviet Union may have been able to foresee the introduction of technical tools that would challenge its mastery on the battlefield at the military strategic level, and it may have simply assumed that the United States would be unwilling to give such advanced technical capabilities to the mujahideen. This experience highlights the dangers of conducting counterinsurgency operations through brutal repression, with the resultant increased likelihood of resistance from the population, if the campaign cannot be effectively continued to its full conclusion. In the case of Afghanistan during the Soviet occupation, the Pakistani safe areas, combined with the technical capabilities of the Stinger missile, guaranteed a defeat of the Soviet intervention in the long term. However, without the introduction of the Stinger missile into Afghanistan, it is quite likely that the mujahideen would have failed to resist the Soviet occupation while the Soviets pursued their successful amoral counterinsurgency strategy.

VI. The United States in Afghanistan

Some obvious lessons can be learned from the Soviet experience in Afghanistan that can and should be applied to the U.S. military intervention in Afghanistan from 2001 to date. It would logically follow that the United States would suffer a similar defeat to that of the Soviet Union if it attempted to conduct a counterinsurgency campaign in the same way. In fact the United States has been accused of repeating the Soviets' mistakes, indicating that the U.S. military intervention is doomed to failure because it is following the same pattern. These commentaries ignore the differences, at times subtle and at times glaring, between the Soviet efforts and the current U.S. effort. Further while lessons can be learned from the Soviet experience, and should be applied, there is no reliable crystal ball for predicting the likely outcome of the U.S. military intervention in Afghanistan.

The most profound difference between the Soviet and U.S. efforts lies with the military capabilities of the Soviet military in the early 1980s and of the U.S. military twenty years later. Arguments that technology does not make a difference in counterinsurgency efforts due to the center of gravity's being in the minds of the population are wholly inadequate.[52] The simple fact is that a military capability that rests on available technical advantages has a direct and controlling impact on the strategic options available to the counterinsurgency forces. As has been pointed out, with sufficient technical capabilities through sensing and targeting, it is possible to implement an operationally offensive, tactically defensive concept to provide maximum protection to small population centers in an effort to separate the population from the insurgents. While such technology may not currently be capable of being fully implemented in the U.S. force structure, it is certainly in the nascent stages.[53] In this context, technology alone gives the U.S. military the strategic flexibility to attempt to win the hearts and minds of remote Afghan population centers. The Soviets could never have implemented this option, even if their military doctrine had called for it.

A further technical disparity between the Soviet military of the 1980s and the U.S. military of the 2010s lies with the latter's ability to use precision-guided munitions in all weather environments to target insurgent forces and strongholds accurately with minimal collateral damage.[54] Again this technical capability minimizes collateral damage to Afghan civilian and logically results in a reduced likelihood of helping insurgent recruitment. Clearly accidents do happen with precision-guided munitions, which may sometimes stray off course and result in civilian casualties; however, the option to use precision-guided munitions allows the U.S. military to attempt to maintain the moral high ground and retain the support of Afghan civilians within an overall strategic context. It is notable that eight years after the initial U.S. military intervention into Afghanistan, 47 percent of the Afghan population still felt that the

United States and its coalition allies were a force for good in the country and not an occupying force with the subsequent negative connotations. Unfortunately that opinion is down from 83 percent in 2005.[55]

The situations of Soviet military personnel and U.S. military personnel in the context of the two military interventions are also notably dissimilar. A typical Russian soldier was vulnerable to small arms fire, lacked a moral high ground, and was not well versed in counterinsurgency strategy. With body armor capable of stopping multiple rounds at close range from AK-47 rifles, U.S. military infantrymen today are almost impervious to small arms fire. While the protection they enjoy is certainly not perfect, leaving extremities vulnerable to attack, the psychological impact of good body armor allows soldiers to act aggressively and confidently when engaged in direct fire action against the enemy. This security on a personal level allows the professional U.S. military soldier to be less fearful and less likely in a hostile environment to overreact, which could result in collateral damage with a negative strategic impact. Such professional capabilities lend themselves to appropriate moralistic behavior, which provides a strong foundation for counterinsurgency strategy in an effort to win the hearts and minds of the local population.[56] This fact can clearly be juxtaposed against the Soviet military experience, which emphasized brutal harshness against the civilian population in order to obtain compliance through fear and thus surrendered the moral high ground to the mujahideen.

Another technical capability that allows the flexibility of counterinsurgency strategy in Afghanistan is the ability of the U.S. military to operate day or night to provide close air support and air weapons teams for hunting insurgents.[57] While there are very efficient man-portable air defense systems (MANPADS), much like the Stinger missile, any country with the technical know-how to manufacture and supply these weapon systems runs the risk of supporting those who perpetrated the September 11 attacks in the United States. Currently no country capable of producing such weapons has a strong ideological bond with the Taliban and has any strategic interest in supplying them to the insurgent forces of the Taliban. Any nation willing to enter the Afghan conflict and pursue such political action would provoke a strong negative reaction, possibly including violence, by the United States.[58] Further, even if a nation that is not ideologically tied to the Taliban chose to introduce MANPADS into Afghanistan for some reason, the U.S. military would retain its air superiority. Moreover it would remain capable of providing all-weather, close air support to ground forces owing to its sophisticated precision-guided munitions. These munitions are incredibly accurate without regard to the altitude from which they are employed and in turn can provide fifteen kilometers of standoff.[59] Soviet close air support in the 1980s was seriously hampered when the Stinger missiles forced the Soviets to move their close air support assets higher than ten thousand to twelve thousand feet above ground level. That system would have little impact on the ability of the U.S. military to provide close air support, which is effective from altitudes well above the range of MANPADS.

These technical capabilities and the military strategic implications should make clear that the U.S. military's counterinsurgency effort has a profoundly different narrative with much greater flexibility than the Soviet experience did in the 1980s. These technical capabilities are currently allowing the U.S. military to attempt to implement counterinsurgency doctrine; however, they will not determine whether the Afghan population, and specifically the Pashtuns in the southern and southeastern areas of Afghanistan, will actually embrace the political ideology of Western democracy and the role of a central government. The technical tools only provide the U.S. military with the ability to attempt to convince the Afghan people that they should embrace a new era for Afghanistan's future. The final choice, though, is up to the Afghan population itself, and technology is unlikely to have a profound impact in such a context at the individual and cultural level.

As of 2009 the strategic context that the U.S. military in Afghanistan faced was also profoundly different from what the Soviet Union confronted in the early 1980s. Pakistan, rather than being a sanctuary for insurgents associated with the Taliban and al-Qaeda, began launching military offensives in the Federally Administered Tribal Areas against Pakistani Taliban organizations. Therefore the Taliban cannot always count on those sectors along the Afghanistan-Pakistan border that were insurgent bases, which provided training, logistical aid, and rest areas, to be secure or to serve as conduits for money and arms flowing from overseas. In the strategic context, the Pakistanis to the south and the U.S. and coalition forces from the north are putting military pressure on the insurgents. The effect is to squeeze the Pashtun areas on both sides of the Afghanistan-Pakistan border, resulting in reduced stability and security in areas that were fundamental to the mujahideen's success in their struggle against the Soviet military in the 1980s. Unlike the Soviets, the U.S. military and the Central Intelligence Agency also have the technical capabilities to use unmanned aerial vehicles to strike targets on the Pakistani side of the border, where insurgents were traditionally secure.

Yet for all the apparent advantages that the U.S. military has for conducting a successful counterinsurgency effort in Afghanistan against Taliban elements, the conventional wisdom in the autumn of 2009 was that the Taliban had been extremely successful in garnering support from the Pashtun populations in southern and southeastern Afghanistan and Pakistan. So while the strategic context and military capabilities are extremely different when comparing the Soviet experience in Afghanistan in the 1980s and the later U.S. experience, some similar trends developed that have forced the United States to contemplate the options of withdrawal, narrative change, or unrestricted operations while still striving to win the hearts and minds of the Afghan population.

Due to the political costs and the loss of prestige both domestically for the Obama administration and internationally for the United States, withdrawing U.S. forces from Afghanistan in 2009 was not seen as a realistic option vis-à-vis the option of changing the conflict's narrative. A withdrawal of U.S. forces from Afghanistan, and a possible resurgence of the Taliban, would have potentially resulted in a political defeat for the Obama administration. It would also have ensured that the United States and its NATO allies could no longer be counted on to enter future counterinsurgency conflicts in support of U.S. political objectives. Further any such withdrawal would have indicated to global jihadist movements such as al-Qaeda that the United States could be defeated anywhere that the civilian population was willing to stand up to the United States for religious and identity reasons. It was logical, therefore, to change the narrative of the conflict in Afghanistan and secure success for the United States on some level.

UK foreign minister David Miliband joined Secretary of State Clinton in 2009 in calling for the incorporation of moderate Taliban organizations into the Afghan government structure. This proposal was based on the concept that the true enemies of the United States were the perpetrators of the September 11 attacks, which clearly were the work of al-Qaeda and not the Taliban. As noted before, President Bush in 2001 had announced al-Qaeda's destruction was the stated objective of the conflict, but mission creep had cast the Taliban as a strategic enemy in Afghanistan. The attempt to change the narrative of the conflict back to its original policy objectives had the benefit of resonating with the American population, who still strongly supported any attempt to apprehend or kill those responsible for 9/11.

Ironically, the attempt in 2009 to change the narrative of the Afghan conflict so as to accept the Taliban as a legitimate political movement was extremely ill timed with regard to Pakistan's strategic position and its fight against Taliban militants in the Federally Administered Tribal Areas. It further risked making the unmanned aerial vehicle attacks in Pakistan, meant to target al-Qaeda leaders and sympathizers, a vulnerable bargaining chip if the Taliban chose to enter into negotiations to become part of the government. Yet clearly any attempt to change the narrative of a conflict due to lackluster results was likely to come at some cost as the opposition—in this case, the Taliban—would likely, and correctly, read such a bid as coming from a position of weakness.

Finally an attempt to change the narrative to incorporate disparate groups of Afghans into a central government structure may be doomed to failure. As the Soviets found out in the 1980s, a strong central government in Afghanistan is not the historical norm, nor are most groups in Afghanistan willing to see local self-determination handicapped and reduced for the unclear advantage of a strong central government.[60] While a strong central government in sovereign control of the geographical region is a comfortable prospect in the Western paradigm of the state system, it may very likely not coincide with the paradigm of Afghan tribal and clan structure. Inevitably concentrating power in a central government presiding over a system of clans and tribes will lessen the power of the tribal structure. There is therefore no reason to be sure that changing the narrative to incorporate the Taliban in a central government will allow the United States to achieve its policy objectives in Afghanistan. And more pertinently technology cannot play any role in aiding the U.S. military as it attempts to achieve Taliban support in the current counterinsurgency strategy employed in Afghanistan. However, if the narrative change fails, technology will become paramount should withdrawal or the use of unrestricted operations becomes required.

It is unlikely that withdrawing the U.S. military from Afghanistan would lead to a total departure of U.S. forces, as had occurred in Vietnam with the subsequent total collapse of all anticommunist elements during ensuing combat operations.[61] It is far more likely that the U.S. military's withdrawal from Afghanistan would resemble the Soviet pullout in 1989, with the United States leaving behind a well-funded and equipped central authority that would probably be doomed.[62] Such a withdrawal would, in fact, maintain both rhetorical and active support for a central governing authority in Afghanistan, though it is unlikely that such an authority would be able to fight a successful counterinsurgency effort without the help of U.S. military forces.

The political proclivity of the United States to support the international system of states, and therefore the notion of a strong central authority in Afghanistan, has hampered the U.S. military's ability to fight a successful counterinsurgency campaign in Afghanistan.[63] The same driving paradigm of the United States is quite likely also to hamper a withdrawal effort. Strategically the United States would be better served in Afghanistan by working in the existing political milieu with independent warlords—an approach that is comfortable for Afghans—and with the aim of separating groups that will respond favorably to U.S. political objectives from those groups that will respond unfavorably. The same advice would hold true in the course of a withdrawal. As Afghanistan is a conglomerate of ethnic groups—most of whom are hostile to the notion of the return of the Pashtun-controlled Taliban—the United States, in the case of a withdrawal, should choose the most successful, or the most likely to be successful, group and support it as a proxy client.[64]

Any chosen proxy client in Afghanistan that is funded by the United States and supported in a limited fashion by U.S. Special Operations

Forces employing technical military advantages, such as precision-guided munitions, would be an incredibly powerful force in Afghanistan even after the U.S. military withdrawal. Just as a limited number of U.S. Special Operations Forces with available precision-guided munitions allowed the Northern Alliance of Afghanistan to successfully beat back the Taliban in 2001, the same would be true for any proxy actor the U.S. military and political leaders chose to support after a U.S. withdrawal.[65] Such a strategy for the face-saving involvement of the U.S. military in Afghanistan would likely serve long-term U.S. interests far better than backing a central authority in Kabul with no actual influence in the country.

Clearly the United States would have to make a decision between a complete withdrawal from Afghanistan or a partial withdrawal while maintaining a proxy actor. It should be equally clear that technology and the military capability gained from that technology's use will play a major part in the viability of any proxy actor in maintaining its influence in the region as was the case following the Soviet withdrawal.[66] If the United States chooses to support a central authority until its inevitable collapse so that the United States could then withdraw completely from Afghanistan while washing its hands of the entire intervention, technology would only play a part. Technology would facilitate an ordered withdrawal of U.S. forces in the face of what would likely be an extremely chaotic situation, with the crumbling of Afghan security forces as they revert to their tribal or ethnic alliances. Such a disorderly situation would have a visual impact, similar to the final withdrawal of U.S. military forces from Vietnam, and would have powerful political repercussions.[67]

VII. Unrestricted Operations

It is extremely difficult to envisage the U.S. military, and especially the civilian leadership in the Obama administration, actually being able to bring itself to sanction unrestricted operations; but they should not now be considered. On a viscerally moral and legal level, such actions would be completely abhorrent to the modern standards of U.S. military personnel.[68] It is also likely the politicians concerned with their legacies and with legality would never think of using unrestricted operations as an option. Many large powers in the world such as China and Russia might ignore such actions if the United States undertook them in pursuit of political objectives in Afghanistan, but Western allies, especially those European nations that attach greater importance to international law, would likely brand unrestricted operations as war crimes or genocide.

More important than international political considerations are the domestic political considerations that would likely have a large, direct impact on the feasibility of employing a counterinsurgency strategy based on unrestricted operations. Given that only 58 percent of the U.S. population rejects the use of torture in a time of war, with the right spin it may be possible, but not guaranteed, to garner enough domestic political support to switch to a counterinsurgency strategy based on unrestricted operations.[69] Obviously a segment of the U.S. population would express strong outrage and indignation to such a sweeping compromise of generally accepted normative behavior, yet norms do not have the same impact as laws. The Guantánamo Bay prison camp, which was established to hold dangerous Taliban and al-Qaeda prisoners without due process on an indefinite basis, violated every normative value in American jurisprudence, but it was not against the law.[70] However, let the author be crystal clear: actions that run contrary to international law should be carefully considered as an option for the U.S. military under certain circumstances. While such a stand is morally and ethically sound, sadly taking a strong moral stand may not be practical under all conditions.

The targeting of Pashtun Afghans who were thought to lend direct support to the Taliban insurgents and their al-Qaeda allies is fundamentally and morally wrong; and this policy in Afghanistan is not far removed from Soviet doctrine in the 1980s or the Russian doctrine in Chechnya in the late 1990s.[71] The list of historical analogies is legion, and any comprehensive survey is quite beyond the scope of this book; however, the well-established historical precedents offer options for militaries to engage in counterinsurgency struggles. In all such cases, and without regard to the actors in question, a moral and normative restriction on the conduct of counterinsurgency operations was a secondary consideration to that of effectiveness. There is clearly no way to predict how the U.S. population would react to a change from the counterinsurgency strategy in Afghanistan to a policy of unrestricted operations. Historical precedents likely indicate that the population will acquiesce to the political leadership and set aside the normative values that Americans hold dear as long as the letter of U.S. law is not violated and an urgent need for such action is demonstrated. However, again let the author be clear: such an option should be carefully considered for the U.S. military.

If the U.S. government deems that a change in the narrative of the war and a withdrawal from the conflict in Afghanistan, as well as continuing to muddle along with a failed counterinsurgency strategy, are considered politically too risky, then it will be out of acceptable options. With the military forces available and the technical capabilities at the U.S. military's disposal, the most pertinent question then becomes whether the U.S. military is capable of conducting unrestricted operations against the Pashtun population of Afghanistan in order to achieve stated U.S. political objectives. The answer is clearly that the United States must only rarely engage in any behavior that is contrary to international law. Many observers of the conflict have openly stated that there is no military solution to be found. In a counterinsurgency struggle, the only effective end state will be achieved through political considerations with the help of military capabilities. While this assertion is true in the context of U.S. counterinsurgency practice, with all its limitations and emphasis on winning the hearts and minds of the people, it is conversely patently false in the context of unrestricted operations.

4   Technology and Counterinsurgency Strategy

> Of course, present-day machines are useful mostly for rather repetitious and somewhat stupid tasks. But there is no reason why they could not become more flexible and versatile, mimicking the intellectual processes of man, with tremendously greater speed and accuracy.
>
> DAVID RUELLE

I. Technology

Technology should never be seen as a substitute for sound strategy. Technology, as applied to counterinsurgency strategy, can potentially further specific military capabilities and is important in the formulation of military strategy in pursuit of political objectives. As a facet of strategy, it is important in that the technology available has a direct impact on the military options accessible to policy makers.[1]

Within the framework of the operationally offensive, tactically defensive concept of counterinsurgency strategy, no single point or set of technical capabilities can suddenly allow the implementation of the concept; rather, a constant dynamic interaction between asymmetrical technical capabilities allows for possible implementation in a counterinsurgency or conventional role. As discussed in chapter 1, the technical symmetry between the North and the South during the American Civil War created a dynamic where only large forces could be expected to implement the operationally offensive, tactically defensive concept effectively. It should be noted that this conventional example is not necessarily applicable to a counterinsurgency situation. As every conflict has multiple variables, actors, goals, and levels of military capability, one historical example may provide insights, but it should not be taken to represent a direct parallel to a different conflict.

Even similar conflicts presenting similar variables will inevitably present different challenges and, very likely, result in different outcomes. The examples of Dien Bien Phu and Khe Sanh should illustrate that the operationally offensive, tactically defensive concept must be examined with each case being considered unique, even when in the same geographical area. Ultimately while every conflict and every battle within a conflict will present specific variables, this book has identified a historical, technical trend toward facilitating the implementation of an operationally offensive, tactically defensive concept.

In the American Civil War, World War I, incursions into Vietnam, and the recent practice of counterinsurgency warfare in Iraq, the historical trend is that greater firepower and technical innovation have led to more capable defensive postures. Imaging, information processing, precision fires, and better force protection have all led to increased defensive capability in modern, technically oriented militaries.[2] This point should not be interpreted as suggesting that all militaries throughout the world have become more effective and lethal while in a defensive posture; rather, the technically oriented and well-funded militaries, such as the U.S. military, are now far more capable in a defensive posture. It is the asymmetry between the technical capabilities of the U.S. military and its likely adversaries that provides strength in the defense.

In a conventional conflict the operationally offensive, tactically defensive concept could have any number of possible applications in the future; however, more pertinent to this work is its application in a counterinsurgency environment. Such an application must marry technical capability with sound strategy in order to achieve desired political objectives. To that end the suitability of an operationally offensive, tactically defensive concept in support of sound strategy will have to be carefully considered on a case-by-case basis.

Also important an operationally offensive, tactically defensive concept in counterinsurgency warfare should not be implemented suddenly in a widespread application unless the milieu is especially suited to it. The inherent risks outlined in chapter 1 will require that any application of the concept be considered only to the limit at which it can reasonably be supported by the integrated technical capabilities in the U.S. force structure. In an effort to secure the populace, the suitability of the concept in counterinsurgency strategy may seem straightforward, with historical technical trends supporting its application. Nevertheless, the technical and contextual realities must dictate the feasibility of the concept in any given conflict.

II. Revolution in Military Affairs

Recognizing the possible increase in military capability by an order of magnitude, the proposed revolution in military affairs is a fairly misleading categorization of the role of technology in military capability.[3] An attempt to define a certain point at which technology has profoundly changed military capability and to define the parameters by which military capability has been changed is neither helpful nor likely to be very accurate. This is not to say that the U.S. military is unable to integrate new technical capabilities to provide greatly enhanced force structure. It is true that technology has radically changed many facets of U.S. force structure with corresponding increases in effectiveness, and they in turn allow greater flexibility for the formulation of strategy by policy makers; however, the integrated technologies did not suddenly spring up overnight to be applied in a revolutionary manner to change military affairs.

An unfortunate tendency in the U.S. military is to see the changing force structure as an entity that existed in one state and will become another. The actual language used by many people in military circles alludes to a propensity to require force structures to be defined as static entities. The term "transformation" itself implies that an entity exists in one state and changes into another state.[4] The U.S. Army's concept of the objective force implies that the current force will somehow be transformed into a force that is an "objective force"—that is, "a full-spectrum, lethal ground force that is 'rapidly deployable and tactically agile' while also conducting 'peacekeeping, humanitarian intervention, and disaster assistance operations.'"[5] Another concept in vogue in the 1990s was the Army after Next, which made an effort to envisage what the U.S. military would be like in 2025.[6] Again it suggested the army would become another future distinct and unique army and would be followed by the Army after Next, which would somehow be a new force. These concepts and words indicate that change is not continuous but rather incremental in nature. The implication is that the U.S. Army or the U.S. military as a whole will somehow reach an end state where military professionals and defense analysts can once again feel comfortable in a static context.

Technology does not lend itself to a static state or to predictable change to facilitate easy definitions and delineations between time periods. If the U.S. military has undergone a revolution in military affairs, it did not start in the early 1990s, and it will not end in the achievement of an objective force. Rather, the technologies that have been integrated into the U.S. military since the early 1990s are, in fact, the products of much earlier technological achievements.[7] The fact is that implementing new technologies that are built upon the development of previous technologies, which themselves were not identified as candidates for acquisition by the U.S. military, does not mean that the process is new. Actually if a revolution in military affairs exists and if it adequately explains the increased technical capabilities of the U.S. military, then it is nothing more than a continuous evolution of technical capabilities as a product of research and development during the past century.

Another deeper problem in trying to define new or improved technical capabilities in a context that describes the U.S. military as a single evolving entity is that the very nature of exponential technical growth may be ignored. According to the U.S. Army, the Future Combat System program, which will be a diverse and autonomous group of ground combat vehicles, will be an integral part of the objective force.[8] However, any notion that a fleet of new combat vehicles, requiring research and development of new technologies, can reach an end state without further development shows a linear thinking about technology. Consider for a moment Brett Steele's categorization of military reengineering in his RAND study of the interwar period:

> *The tabula rasa is a myth.* I can state, categorically, that in none of the interwar reengineerings did the military leaders start with a blank sheet of paper. Although Tukhachevsky appears to have come close, his reengineering effort applied to only a relatively small section of the Red Army. As with the Wehrmacht, the vast majority of Russian troops continued to

maneuver themselves tactically by marching. These armies never engaged in a sudden, radical transformation. Instead, they preferred to pursue evolutionary strategies accompanied by intensive study, debate, testing, and adjustment over lengthy periods of time. However impressive the Wehrmacht was in organizing its panzer divisions in 1935 and subjecting them to combat by 1939, its success rested on many years of small-scale research, training, and observation during the Weimar Republic and early Third Reich.[9]

Even during this period of dynamic change, with the integration of the internal combustion engine, sudden transformations clearly did not take place. Concepts, technology, and the implementation of new methods to employ the new capabilities successfully were the result of evolutionary rather than revolutionary change. The U.S. Army's current revolution in military affairs is a similar process. It is very unlikely that a fixed set of technologies will transform themselves into new force structures with a specific and static end state. The historical narrative is one of continuous evolution in research and development in response to new possibilities provided by new technologies.

Rather than viewing it as a straight line of easily defined, ever-increasing capabilities, such as the increased range and lethality of a tank's main gun, technology should be seen as a tree with ever more branches. This fractal and chaotic representation of the exponential growth in the possibilities of technology captures the ever-increasing complexity as different technologies build upon each other to provide new possibilities and new applications. Consider the requirement that the National Aeronautics and Space Administration (NASA) had for a cordless drill on the moon during the Apollo missions in the 1960s and early 1970s; later it led to the development of precision cordless surgical tools and miniature vacuum cleaners.[10] Every year NASA features the top fifty spin-off applications of NASA-developed technology that have spawned advancements outside of NASA.[11] Cell phones, laptop computers, a plethora of cordless power tools, communications devices, navigational aids, and many more similar items can all directly link their development to NASA, which has funded battery research since the 1960s.[12] Of these items, now consider the applications that have spawned from the laptop computer, from producing a doctoral thesis to uploading GPS coordinates directly into munitions on a distant battlefield. The potentials are almost limitless. To list all the spin-off technologies and capabilities that have evolved from this one tree of technology would be next to impossible. It is even more difficult to conceptualize how these technologies have combined with other technical trees to provide new possibilities and new military capabilities. For example, the components that go into an M1 Abrams tank's main gun system—from the metallurgy of the actual barrel to the computer processing of its fire control system, to the optical systems and laser range finding, to the innumerable technologies that provide mobility to the gun, and to the integration and communications required for target acquisition—represent technical trees that have been built upon over decades and will continue to be refined in future decades to provide unknown new capabilities.[13]

A further example of a product building on multiple technologies is that of the modern car. Consider the following excerpt from a BBC report marking the fiftieth anniversary of the invention of the laser: "'There is a phenomenal amount of laser processing on a car; you wouldn't believe how much—laser cutting, marking, measurement, drilling, hardening, laser brazing, laser deposition, and laser welding,' explained Tim Holt, the chief executive of the Institute of Photonics, University of Strathclyde. 'Modern cars today would not be possible without lasers.'"[14] In essence the modern car is not simply an evolutionary descendant of pistons and gears but an amalgamation of technical innovations and advancements in research and development from many technical trees to produce a vehicle in a wholly different manner. The function of the vehicle and its purpose may have remained closely aligned with the original concept of the automobile, but the industrial ability to make a modern car relies on a suite of different technologies that have been developed and tested slowly over time.[15] Could it reasonably be argued that there has been a revolution in automotive affairs through the integration of computer technology and laser manufacturing into the production of cars? Should any attempt be made to date this potential revolution in automotive affairs, and if so, does it truly have any meaning? The reality is that cars may appear to have taken a revolutionary leap since the 1960s because of the technology incorporated in them; however, change has been slow and gradual as new technologies have combined in more productive industrial applications that have allowed the manufacture of a better car. Its evolution has not been sudden, and it should be expected that cars will continue to evolve as new and novel technical applications for better production become a reality. There should be no expectation of a "car after next" or an "objective car" or any serious plans by an automobile manufacturing company to produce a transformational car that will totally change how people drive. The notion of an objective car sounds absurd, yet, strangely, an objective force can resonate through defense circles as forward-thinking wisdom.[16]

Thus to classify the possible increase in military capability as a revolution in military affairs, and to date such a process with a beginning and, more inanely, an end, is to ignore the true nature of technology and its compounding, or the exponential growth in the absolutely limitless possibility for new combinations to provide new capabilities.

This chaotic and compounding effect highlights a common misconception of technical transfer between actual or potential adversaries. During World War I the army provided the Browning automatic rifle in limited numbers to the U.S. infantry fighting in Europe.[17] The thinking at the time was that the weapon was too advanced and qualitatively so far ahead of German weapons that it should not be used or it would likely be captured and replicated.[18] Beyond the baffling notion that only inferior weapons should be used in warfare, this example highlights the perception that technological transfer, or creep, tends to negate the technical superiority or advantage of one party vis-à-vis another party. This observation is probably accurate when dealing with an item as simple as an automatic rifle, as undoubtedly Germany would have been perfectly capable of replicating and manufacturing a Browning automatic rifle in 1918. However, the likelihood that any competitor could replicate a modern, highly complex military system, such as the F-22 Raptor, would be nearly impossible within the time frame of any armed conflict in which a state actor could engage the United States. The technical complexity of such an advanced aircraft is not the product of a few high-tech components; rather, it is the product of generations of technology improving upon each other in both the aircraft itself and the industrial processes that more than twenty-seven highly specialized defense contractors use to manufacture it.[19] To replicate a Raptor would require not only the capability to reverse engineer the weapon system but also the incremental and evolutionary development of the technologies even to begin to develop the industrial techniques to produce the system and its subsystems.

It may be reasonable over the course of the Cold War to have expected a leakage of U.S. technical capabilities to the Soviet Union that in turn could be replicated, such as the B-29 bomber after World War II.[20] However, that case is far different from a hypothetical situation in which an F-22 Raptor is shot down by Iran, reverse engineered, put into production, and then fielded against U.S. forces before the U.S. military could destroy Iran's industrial capacity.[21] The more technically complex a platform is, the harder it is for the technology behind it to be disseminated for exploitation against U.S. military forces. This is not to say that technology will not be disseminated to likely adversaries of the U.S. military; rather, as the U.S. military continues to far outspend any other state, or non-state actor, in research and development of new technologies and new applications that lead to new capabilities, the technologic advantage of the United States will likely remain significant.[22]

The proposed revolution in military affairs is neither new nor linear in nature, and any attempt to strictly define it, either chronologically or in its technical scope, becomes a meaningless endeavor and actually detracts from the potential of technology. Technology should be accepted as ever flowing, as dynamic, and as an unpredictable combination and recombination of possibilities; it then becomes feasible to conceive of capabilities outside current doctrine and force structures. Rather than an evolutionary process by which the military must develop better tank guns to place on more sophisticated tanks for improved tank performance in armored divisions, it becomes possible to work outside the established paradigm.

Take, for example, the line-of-sight antitank weapon, which can be mounted on the top of a Humvee or any other lightly armored vehicle.[23] The system is capable of driving a long metal rod penetrator through the front of the most advanced battle tank in the world, and it can do so beyond the range of any main tank gun round.[24] Further, the logistical support, ease of transport, and manpower requirements for the system are far lower than that of a tank. Beyond the obvious ability to penetrate armor, the important point of the system is that it brings together the technologies behind solid rocket fuel, and with extremely fast computing power, it relies on information rather than composite armor for protection.[25] This system is a prime example of multiple technologies not generally associated with tanks coming together to provide a new and innovative military capability that is completely outside the conventional paradigm. The line-of-sight antitank weapon, while being a conventional system, is an example of radical thinking transferred from one military capability to another rather than the evolutionary technical development of a single capability.

Similarly the revolution in military affairs would be more aptly described as a continuing integration of novel technology to improve the military's capability. The process is neither new nor revolutionary but has the potential to provide radical new capabilities when properly focused to support strategy.

III. Technology in Counterinsurgency

One of the most intractable problems in applying technology from a conventional force to facilitate a counterinsurgency strategy is the misapplication of technology in pursuit of an unsound strategy. The U.S. military faces a conundrum in that it must always be ready to fight against a peer competitor in a conventional conflict; yet, at the same time, since 2001 circumstances have demanded that it implement a counterinsurgency strategy. As Steele noted, any reengineering effort will not quickly or transformationally do away with legacy forces and force structures; therefore, it is understandable that the U.S. military has attempted to use some of these legacy capabilities where they may not be most applicable. Donald Rumsfeld noted during the invasion of Iraq that a nation must go to war with the military it has and not necessarily the military that it wants.[26]

The root of the problem in trying to tailor technology for counterinsurgency operations is that the technical capabilities that provide for the most efficient and lethal conventional force do not necessarily translate into the best technical capabilities for waging a counterinsurgency effort. Conversely the best technologies and applications of the technologies that lend themselves to the most efficient and effective capabilities for a counterinsurgency effort are not always the most suitable technologies for a conventional conflict. While the perfect solution would be for the U.S. military to have a force structure capable of waging both conventional and counterinsurgency conflicts, the political and financial price of funding two separate force structures to fight two different styles of conflict is not reasonable. Given that it can take twenty years from the conception of a military need to the actual fielding of a military capability, it is hard for the U.S. military to justify changing its force structure from conventional to counterinsurgency for a conflict that will run for an undetermined amount of time.[27] Obviously if the U.S. military were to change the conventional force structure to a counterinsurgency force structure over the course of twenty years, then the United States would have to examine whether it would be prepared to engage in a conflict with a peer competitor such as China or Russia. If it made the decision to change the force structure over the course of twenty years and the U.S. military then found itself in a situation where it had to consider a conflict with the peer competitor, it could take another twenty years for the force structure to revert to a conventional posture.

Neither changing the force structure of the U.S. military nor creating two independent force structures to fight different types of conflicts is reasonable or politically viable. The only option open to the U.S. military is to develop technologies that provide both conventional and counter-insurgency capabilities for the same force structure. This dual capacity would allow maximum flexibility in military capability and, therefore, strategic flexibility for policy makers. To be effective it is therefore most appropriate and necessary that research and development projects produce technical capabilities that are directly targeted to be operable in either type of conflict.

It can be argued that two force structures do, in fact, exist in the U.S. Army to fulfill the conventional and the counterinsurgency roles that may be demanded of it. The conventional U.S. Army is expected to fight against peer competitors in conventional conflicts.[28] Augmenting this force structure are special operations forces, specifically U.S. Army Special Forces, that are capable of conducting unconventional or irregular war-fare, of which counterinsurgency is an aspect.[29] However, the prospect of relying on special forces alone is illusion rather than reality. Consider the stated political goal in Iraq—to develop and protect a Western-style democratically elected government and its continued viability.[30] While special operations units are fully capable of conducting raids and the U.S. Army Special Forces of building indigenous fighting units, neither alone are able to fulfill the political objective of instituting Western-style democratic governments in either Iraq or Afghanistan. Political endeavors of this magnitude require so many troops that the conventional units must take part and more likely constitute the majority of the effort to ensure the sustainability of the newly formed central government. Special operations forces simply do not have the manpower to be able to transform a state's political system in the international system of states.

U.S. Army Special Forces also lack the manpower to embed themselves with an indigenous culture, such as in the Pashtun areas of Afghanistan, on a scale large enough to give personnel effective control of the countryside. Post 9/11 special operations forces have moved away from the traditional role of working by, with, and through indigenous peoples.[31] This drift toward less interaction will likely change in the future, and special operations forces will once again focus more on people skills rather than rely so heavily on its kinetic capabilities. In ceding the countryside to insurgents, the U.S. military left a large segment of the population unprotected and open to exploitation. The only pool of manpower large enough to theoretically protect rural population centers from insurgent control and exploitation is the U.S. military's conventional forces. Coalition allies and host nation forces cannot be relied on to fulfill such a role as these units will inevitably find themselves in combat and taking a high number of casualties with the attendant political costs.

The conventional U.S. Army should not be expected to perform the role currently filled by U.S. Army Special Forces teams. Most conventional units lack expertise and training to effectively become part of indigenous societies, build guerrilla bands, and secure the local populace from insurgent forces. After a rigorous selection process, U.S. Army Special Forces soldiers typically undergo more than a year of training to ensure they can not only survive but also thrive in such an environment.[32] To demand that conventional U.S. soldiers achieve the same level of maturity and expertise for counterinsurgency operations, at first glance, may seem an unrealistic expectation; however, what conventional soldiers lack in expertise and training can be partially made up through the use of technology.

Personnel in a typical U.S. Army Special Forces Operational Detachment Alpha must embed themselves in a local society and live among the people to earn their trust and build effective indigenous fighting forces.[33] While security is always an issue for a special forces team, which must take steps to minimize its security risks, an overabundance of force protection to the detriment of the mission is not considered an option.[34] A special forces team does not erect a compound out and away from the indigenous population and minimize contact with the population while, at the same time, expecting to earn their trust. It is imperative that the team intermingle with the local population while helping the people in every possible way in order to garner trust. A special forces team often takes care of villagers' medical problems, veterinary issues, and construction and building matters, as well as train the local force to be an effective militia.[35] Making daily contact and building personal relationships allow a special forces team to establish a solid rapport with the indigenous people and convince them to support U.S. policy objectives.

Since effective large-scale counterinsurgency operations will require the participation of U.S. conventional forces, it becomes necessary to identify and foster the technologies that will allow conventional forces to mimic the actions of the U.S. Army Special Forces ODAs. The proposed revolution in military affairs, which is simply a compounding exponential growth of technology, needs to be tailored to develop the technologies that will allow conventional forces to supplant A teams while not losing or destroying the ability of the conventional forces to fight a peer competitor. This requirement may seem an extremely tall order to achieve; however, it is exactly how the U.S. military should be harnessing the exponential growth in technology. Precisely this kind of enhancement of military capability, and the pursuit of sound conventional or counterinsurgency strategies, represents the potential offered by the technical innovations of the past century.

Many technologies being developed for the U.S. military force structure do, in fact, lend themselves to both the conventional and the counterinsurgency realms of conflict. Most notable is the idea of net-centric warfare in which information and the sharing of information become enablers for military capability.[36] It is a generally accepted idea that information leading to situational awareness is an invaluable asset to military commanders on the battlefield.[37] The ability to share information between different echelons, or different nodes within a network, will provide a "lifting of the fog of war" and result in unparalleled coordination in targeting and destroying enemy forces.[38] While it might not be self-evident, the same technology is at the core of a sound operationally offensive, tactically defensive concept. In this case, the technology transcends the classification of warfare, between conventional and counterinsurgency, and is properly considered dual use.

IV. Technology and Current Counterinsurgency Strategy

Many legacy technologies used in Iraq and Afghanistan, it can be claimed, are a part of a sound counterinsurgency strategy. Sometimes these legacy systems work very well in supporting counterinsurgency strategy, but at other times the technology is wholly inadequate, inappropriate, and sometimes even counterproductive. As stated previously, the misapplication of technology is one of the biggest problems in actually conducting counterinsurgency strategy; it is eclipsed only by poor strategy itself. It should be recognized, however, that the U.S. military must use the tools on hand to achieve its missions in pursuit of policy objectives.

The mechanized mentality of the U.S. Army has a tendency to hamper effective counterinsurgency strategy. In Iraq during the super forward operating base era between 2004 and 2006, U.S. forces developed a tendency to commute to a fight; that is, U.S. forces were in large bases in a defensive posture and would then drive out daily to patrol areas where insurgents were thought to be.[39] The net effect was a continuous stream of improvised explosive device attacks on U.S. military vehicles and a separation of the U.S. military from the population it should have been protecting. This example illustrates both of the problems listed earlier: poor strategy and the misapplication of technology in counterinsurgency conflict where mechanized forces used their vehicles to separate themselves from the population.

The average infantryman in the U.S. Army today has more kit than he is able to carry.[40] Most American soldiers literally cannot pick their gear up and move with it. U.S. Army tankers also have considerable logistical supply requirements. The logistical needs of armor units also require men and matériel to support the logistical needs of the support units. The logistical tail of the U.S. Army as a conventional force, by necessity, has driven the force to be a road-bound army tied to predictable and long logistical lines.[41] Even in Iraq by 2008, with the new counterinsurgency strategy outlined by General Petraeus, it was not uncommon for soldiers to own laptops, gaming consoles, and televisions, all of which were readily available for purchase in base post exchanges.[42] While such amenities provide a boost to troop morale, they also tend to make soldiers more sedentary and less mobile. It ties the soldiers—or it is at least an indication of how tied a soldier can become—to an insular existence, which runs against sound counterinsurgency strategy.[43]

In one incident near FOB Caldwell in early March 2008, soldiers from the Third Squadron, Second Cavalry Regiment, Fourth Brigade, Second Infantry Division, moved out on a patrol to set up a roadblock several miles from the base. The patrol consisted of two Humvee vehicles and two M1A2 main battle tanks. On the way to the roadblock site, the convoy stopped in a small village and questioned and fingerprinted a local merchant who happened to be writing something on a piece of paper as the convoy passed. When the convoy stopped, children on their way home from school walked past the rumbling massive forms of the tanks. The children were obviously disturbed by their presence and the sounds as traffic came to a standstill and no vehicle was allowed to approach the convoy. The displeasure of the locals was evident. After about twenty minutes, the vehicles moved on and set up a roadblock in order to search Iraqi vehicles and, theoretically, apprehend insurgents.[44]

The brigade was generally very successful and performed counterinsurgency operations according to sound strategy, and this example is a glaring exception. First and foremost the soldiers commuted to the roadblock through a town in which they scared the local population, especially the children; interrogated a local who was later determined to be doing nothing wrong; and interrupted people's daily lives. All of the negative effects stemmed from the fact that two battle tanks were involved in the patrol. The soldiers on patrol were more interested in and concerned with force protection than they were with how the local population would perceive them. Obviously there is a time to impress upon locals the technical prowess and firepower available to the U.S. military; however, this village was not more than a mile and a half from FOB Caldwell, which the soldiers had occupied for the better part of a year.[45]

The roadblock was counterproductive and a further exercise in futility. In short order a line of about fifteen cars backed up on the road from both directions. Vehicles were searched and allowed through the roadblock one at a time. Any insurgent, even one carrying contraband materials such as explosives, could have easily seen the hulking tanks that established the roadblock and simply turned around instead of queuing with the other cars. So in short, the only people inconvenienced that day were ordinary citizens who were going about their business but had to queue for forty-five minutes on an empty stretch of road and then endure searches of their vehicles and their persons, which was culturally insensitive.[46]

After about two hours, having completed its mission, the patrol headed back to FOB Caldwell. As an end result, this patrol consumed hundreds of gallons of fuel, put wear and tear on the military vehicles, endangered American soldiers in a meaningless roadblock with the potential for becoming the target of a vehicle-borne IED, scared local schoolchildren, disrupted people's daily lives, and saw the further separation of U.S. military personnel from the people the post was supposed to be protecting. This example shows not only a clear lack of sound counterinsurgency strategy but, more important, with its use of heavy armor in a routine patrol, a misapplication of technology to counterinsurgency practice.

Unfortunately force protection, which is the impetus for using tanks on regular patrols, is a serious enemy of sound counterinsurgency practice. Force protection is also a highly technical aspect of the U.S. military. Casualty intolerance is a by-product of continuous news coverage and generates such high political cost that it inevitably drives the U.S. military to seek to limit the number of casualties at all times. The U.S. military should not throw away the lives of soldiers, but it should be more focused on accomplishing the mission rather than on protecting the lives of soldiers as a matter of political expediency.

The mine-resistant ambush-protected vehicle is another befuddling misapplication of technology to counterinsurgency operations. The vehicle itself weighs more than forty-eight tons, half as much again as a World War II–era Sherman tank, and stands more than twelve feet off the ground.[47] Its height is a result of the incorporated V-shaped underside that provides standoff from IEDs and helps to deflect blast energy away from the troop compartment.[48] The MRAP is undoubtedly the safest vehicle to travel in on roads where IEDs are thought or known to be; however, the vehicle's drawbacks are as spectacular as the force protection it offers. The vehicle is so tall that it cannot be driven in many smaller villages because it exceeds the height of many power lines strung across streets. Tearing down power lines does not endear the U.S. military to local populations. More important, the MRAP has a tendency to compromise strategy.

In March 2008 A Battery, Second Battalion, Twelfth Field Artillery Regiment, Fourth Brigade, Second Infantry Division, sent a patrol from Combat Operating Base DMC south to the village of Abu Khamis. The purpose of the patrol was to project a military presence and to ensure the locally recruited militia was keeping al-Qaeda from reentering the village and intimidating the locals. COB DMC was located only three to four miles north of Abu Khamis. The patrol mounted up in MRAPs to drive the few miles to Abu Khamis, then it engaged in an almost comical effort to turn the vehicles around in the narrow streets for the return trip. Soldiers who had been observed operating very effectively in air assault missions piled out of the MRAPs to begin walking through the village while exposed to possible IEDs.[49]

While the MRAP is well suited to road movement and offers protection, the Abu Khamis example shows it has detrimental effects as well. A far more appropriate tactical movement would have been achieved if the group had walked to the village from the east or west at night. Stopping outside the village beyond small arms range would have offered the group a chance to observe the local militiamen at first light, when they would have been unaware that they were being observed. American soldiers entering the village would have offered little or no time for insurgents to escape unnoticed by a supporting unmanned aerial vehicle. The locals would have realized that the U.S. forces could appear at any time and from any direction while the U.S. forces would have had the opportunity to mingle suddenly and unexpectedly with the locals to gather information and assess the village. Word would also have leaked to al-Qaeda that the American soldiers at the DMC were unpredictable. The MRAP and its protection offered an easy and relatively safe ride to the village, but it made the patrol far less effective than it should have been from a tactical standpoint in harmony with sound strategy.

The effective technologies employed in pursuit of counterinsurgency strategy are more often centered on information rather than on legacy-oriented widgets. Understanding a battle space in the conventional realm is not wholly different from processing information in the counterinsurgency realm of warfare. Situational awareness offered by ground surveillance radars or air- and space-based radars can be of profound importance in both the conventional and counterinsurgency environments. The Joint Surveillance and Target Attack Radar System is a prime example of a platform capable of serving a dual role in providing information about ground movement.[50] The classification and tracking of vehicles have obvious applications in a conventional setting and in the counterinsurgency realm, even though it is less effective in the latter role.

Precision-guided munitions and platforms capable of precision strikes such as the AC-130 Spectre Gunship allow for timely, appropriate, and discriminate targeting in a counterinsurgency role.[51] Space-based sensing and observation allow the collection of information for targeting and analysis, which also benefits a counterinsurgency effort.[52] In short the U.S. military uses many systems today that fit an appropriate tactical role and enhance a sound counterinsurgency strategy; however, information systems and air- and space-based assets do not separate insurgents from populations. So while the information and kinetic systems currently employed can effectively support a sound counterinsurgency strategy, they cannot substitute for one.

The one unambiguously dual technology currently in use that is indispensable in the counterinsurgency realm is the modern body armor worn by American infantrymen.[53] As discussed in chapter 2, body armor has significantly reduced the wearer's likelihood of fatality from small arms fire, but American infantrymen are not impervious to small arms fire. In any firefight if enough rounds are fired at soldiers protected by body armor, then inevitably they will eventually be wounded, sometimes fatally.

Body armor's protection from normally catastrophic wounds, however, enhances the wearer's confidence and morale.[54] It also offers survivability by allowing time for help to arrive via close air support or reinforcements. More important, as there is no realistic "objective force," there is no final version of body armor as well. Materials science will continue to explore the possibilities of stronger fibers, different weaving techniques, and even reactive armors to continue to enhance survivability.[55] Body armor is truly a technology that offers enhanced tactical capability in support of sound counterinsurgency strategy. The tactical survivability of American soldiers translates directly into an ability of the infantry unit to protect a population from insurgent intimidation and exploitation.

V. Sensor Technology

The inherently political struggle that underlies any counterinsurgency effort requires the host nation to be able to secure enough of the population to retain legitimacy and work effectively against the insurgents' force structure and the political grievances that gave rise to the insurgency.[56] Any U.S. military effort to support the host nation should therefore be concentrated on the security of the populace to allow the host nation to trade time for political reform. This political process also requires that the U.S. military tactically trade time for space in its effort to secure the populace from insurgent influence. Trading time for space becomes crucial as a tactical component of the operationally offensive, tactically defensive doctrine as it effectively harnesses technology to compensate for troop dispersal.

Sensor technology facilitates the deployment of a sensor network, and the utility it offers is not strictly limited to a specific population or geographic setting. "A sensor network is a network of many smart devices, called nodes, which are spatially distributed in order to perform an application-oriented global task. The primary component of the *network* is the sensor, essential for monitoring real-world physical conditions or variables such as temperature, humidity, presence, sound, intensity, vibration, pressure, motion and pollutants, among others at different locations."[57]

Any environment offers the possibility of using sensing technology to differentiate phenomena that could be used to provide the intelligence that should be at the heart of any counterinsurgency effort. Urban areas are no different from rural areas in providing sensing opportunities. Consider San Francisco, which has a networked node of acoustic sensors to identify and triangulate the sound of gunfire to improve the reaction time of law enforcement officials.[58] Airports are man-made choke points that allow sensors such as biometric and full-body scanning to screen large numbers of individuals.[59] Biometric data also can be gleaned from automatic teller machines and traffic cameras to help locate individuals of concern.[60] Hospitals can record genetic data, while sewer systems can be used for measurement and signature intelligence analysis of neighborhoods where illegal substances such as explosives may be mixed.[61] It could well be argued that urban areas provide greater sensing opportunities than rural areas do, as well as a greater likelihood of forming sympathetic militias.

An example of a counterinsurgency effort applied in an urban environment in which the insurgents were tracked and broken was demonstrated by the French in the battle of Algiers.[62] Such ruthless measures are not currently viable for the U.S. military, but the utility of sensor technology can help to make up for that deficiency if it is linked to strong HUMINT efforts and constructive engagement of the local populations as dictated by sound counterinsurgency practice. Fortunately for the host nation, centralized urban populations are not typically viable, long-term territorial bases for an insurgency. Insurgents will not usually control the urban environment to the extent that they can use it as a base area for rest, training, and recruitment. This observation does not hold true for an urban area that is entirely situated in insurgent-controlled territory and devoid of effective governmental structures or an authoritative presence, as was the case in Kandahar, Afghanistan, or Mosul, Iraq, in 2009 and possible coastal megacities of the future.[63]

Rural areas are particularly significant. The settlements tend to be scattered, which is dangerous for counterinsurgency forces, and closer to insurgent base areas, where the insurgents' presence is stronger. During their respective Vietnam conflicts, both France and the United States were forced to cede rural areas, where the communists were given free rein to control and recruit.[64] As the operationally offensive, tactically defensive concept for counterinsurgency stipulates, forces should focus on the rural areas and secure their small populations to prevent their use by insurgents as base areas. Sensing becomes a critical technical component to ensure the concept is viable without risking defeat in detail.

Sensing is a relative term that can denote an entire spectrum of fidelity and usefulness. For example, after the battle of Tora Bora, the infrared (IR) signatures of campfires were detected along probable egress routes of escaping Taliban and al-Qaeda fighters. U.S. forces did not attack them with air-delivered ordnance, however, because there was no way to determine if they were from fleeing fighters or shepherds camping with their flocks.[65] Simply receiving an event signature does not automatically translate into useful and actionable intelligence. In fact, it can be detrimental if it tends to overload users with large amounts of meaningless data.

Consider a hypothetical scenario in which overlapping sensors are collecting information or looking for anomalies in a given area. An IR sensor may detect a heat signature in a mountainous area of Afghanistan thought to be used as a transit area for Taliban or al-Qaeda fighters. The heat source could be an insurgent fighter or perhaps a wandering bovine or even a couple of goats huddling together for warmth. To add a better understanding, a space-based radio frequency (RF) sensor looks at the area and determines that an electronic device is in the same area as the IR energy source. An acoustic sensor then scans the area and determines that a human is speaking, ruling out the presence of animals. Simultaneously, chemical sensors downwind from the location start picking up traces of gun oil and cordite. In this situation, the targeting of the insurgent may be justified based on overlapping sensor information. Therefore, an unmanned combat air vehicle (UCAV) is called in and broadcasts a video feed that shows two men carrying weapons. The determination is made to attack the targets, and they are killed. In this scenario, sensors may have appeared to perform a valuable intelligence service. A subsequent patrol by coalition forces would actually find two dead shepherds, with rifles to protect their flocks, and a portable radio tuned to the *Afghan Star*'s broadcasts.

In this example, the sensor capabilities described are all actual tools available to the U.S. military. The hypothetical sensors performed as intended and provided an overlapping picture with a degree of fidelity, but this level of sensing simply detects a phenomenon instead of classifying the phenomenon in measurable terms that allow meaningful insight. It does not mean that the sensing or the ability of sensors to detect an event without measuring it is a handicap; rather, sensors need to be able to process data or be linked to a network capable of processing data for their full benefit to be realized.

Forward-looking infrared sensors on such air assets as the AH-64 Apache helicopter or ground-mounted observation devices are extremely valuable assets for static defense and kinetic operations in cold ambient environments.[66] However, the thermal images that they produce require a human observer to interpret their meaning and assess the relevant value of the information. In effect, the requirement for a "man in the loop" will not change immediately, but the information that the sensor provides can be enhanced.[67]

Any phenomenon that can be detected can be classified or measured with greater fidelity. The issues with classification or analysis of phenomena are generally engineering or computing matters. Consider the acoustic sensors in San Francisco that can triangulate the location of gunshots fired. The microphones are calibrated to detect sound in a certain range and autonomously classify it as gunfire. The same technology can be taken further, provided the time and monies are available, for enhancing the capabilities of the sensors and/or the processing of the sounds that the sensors identify. Algorithms have been developed as an evolution of earlier voice recognition sensing to detect speech characteristics that help identify a person's nationality.[68] Their purpose is to help spot individuals who may be trying to hide their origins or nationalities when entering the United States. With the proper sampling and processing power, the sensors could identify a person as an Arabic speaker as efficiently as a sufficiently trained security officer can. In addition beyond simple classification as an Arabic-speaking individual, the person's speech patterns and pronunciation could be discerned as belonging to a particular geographical region, such as an area near Mecca, Saudi Arabia.[69] The ability to accurately identify a person's spoken language and to determine the geographical region the individual is from would have prevented the hypothetical mishap with the shepherds. If the acoustic sensor detected the voices to be Tajik, Uzbek, or another non-Pashtun language, then it may have given the UCAV operators reason to pause. But if the language detected was recognized as Yemeni Arabic, Saudi Arabic, or another Arab dialect, then its finding would have lent weight to the assumption that the individuals were foreign insurgents.

Likewise, detecting an RF signature is far different from classifying the RF signature. The ignition in every vehicle ever manufactured, provided that it is electrical, has a unique signature, like a fingerprint.[70] A satellite phone and a small portable radio, for instance, have different signatures. Detecting an RF energy source in the mountains of Afghanistan, while helpful, is not as important as classifying the RF phenomenon to identify the type of electronic device emitting it. In the shepherd example, the implication of detecting a portable radio is far different from that of a satellite phone. Clearly the average Pashtun shepherd is not tending his flock while using a satellite phone to make calls. If the presence of a satellite phone was matched to the acoustic signature of a Yemeni Arab, the case for a kinetic strike would become far stronger.

A current problem with the U.S. military's ability to remote sense is a *soda straw effect*, which refers to the view obtained when looking through a drinking straw: the very small area viewed through the straw is clear, but everything outside that area is unknown. The ability to understand an environment in detail is well developed with intelligence, surveillance, and reconnaissance (ISR) assets; however, that ability is compromised as the area under scrutiny is expanded. So while it is possible to understand fairly well what is happening, where people are moving, and the likely intentions of insurgents in a given area, that level of fidelity is only temporary as assets are inevitably needed elsewhere.[71] If sensors are going to play their appropriate role in U.S. counterinsurgency operations, the soda straw effect must be reduced or eliminated.

VI. Technology and Air Operations

As kinetic operations are too often focused on in counterinsurgency efforts, the same lack of appropriate emphasis can happen when examining airpower. While kinetic air operations are important, even indispensable, non-kinetic air operations are of equal value, especially if the operationally offensive, tactically defensive concept is to be employed.[72] The role of airpower in counterinsurgency operations has been well examined and certain general conclusions can be reached; however, new technologies have been implemented that should alter conventional wisdom.

The siege at Dien Bien Phu could be cited as an example of why aerial resupply of a surrounded garrison is not a sound tactical course of action. The same argument applied to Khe Sanh would be much less valid as massive close air support provided enough standoff from enemy forces to allow aerial resupply. In both battles, it was rightly assumed that effective aerial resupply meant that aircraft would have to land to unload the supplies for the garrison. Logically it would follow that aerial resupply requires a runway and enough standoff from enemy positions to allow effective resupply to happen. This is no longer true.

Parachuting supplies into a garrison to ensure its survival is a valid way to resupply material. As was seen at Dien Bien Phu, this method became exponentially less effective the smaller the garrison became as some pallets inevitably dropped short or long of the intended point of impact.[73] Technology completely changes the likelihood of an accurate resupply drop inside the area controlled by the garrison. By attaching a global positioning satellite guidance system to a parafoil, the resupply pallet, weighing up to twenty-eight thousand pounds, can be dropped from twenty thousand feet above mean sea level and be assured of landing within a hundred meters of the desired impact point after exiting the aircraft from up to nine miles away.[74] As with the Fulton surface-to-air recovery system, the same delivery system could be equipped with a balloon to send the guidance and foil components back aloft for retrieval by a passing aircraft.[75] The GPS and recovery components could then be repackaged for reuse. Garrisons can now be supplied by large transport aircraft ejecting pre-packed pallets with impunity at high altitude well beyond small arms or shoulder-fired surface-to-air missile range and with an extremely high degree of accuracy.

While supplying and retrieving supply packets could easily be accomplished from high altitude, medical evacuations still clearly require rotary wing aircraft to land, presenting a potential for losing aircraft and aircrews. The best defense against such a threat is offered by ISR and close air support assets to identify and suppress insurgent activity while a medical evacuation occurs. Current medical evacuations result in soldiers receiving medical care on average within twenty minutes from the time of a wound or injury.[76] Supporting air assets are the only reason why soldiers can receive such prompt medical attention.[77] As a result of prompt medical evacuation and fast, quality trauma care, a soldier in Iraq had less than a 12 percent chance of dying from a wound, down from greater than 24 percent during World War II.[78] As such, air mobility and the technologies that make it as fast as possible correlate in reduced fatalities with their attendant political costs.

The current force structure of the U.S. military makes it impossible to transport all matériel and personnel by air. The only way to move the equipment required to sustain or deploy units is by ground transport, which must move along predictable lines of communication. These routes are prone to IED emplacement or ambush by insurgent forces. IEDs accounted for the majority of casualties suffered in Iraq as they allow insurgents to inflict casualties on U.S. forces with little exposure and risk to themselves.[79] The predictable movement of supplies on known routes offers easy targets for insurgents to interdict or capture supplies and force the counterinsurgency forces to spend capital or manpower to protect them.[80] The use of air transport would negate all of the problems associated with ground transport along lines of communication; however, the U.S. military would have to dramatically rethink its force structure and funding priorities before a campaign could rely solely on air assets for logistical support.

UAVs and UCAVs have tremendous potential for radically changing the availability of air assets, though often a shortage of pilots rather than a shortage of aircraft is an issue.[81] UAVs have the advantage of limiting the potential political costs of a counterinsurgency effort; dead soldiers can elicit a negative, visceral response from the American public. While UAVs are not generally designed to be expendable, there is very little political cost if they are lost during combat operations. Provided enough bandwidth is available to control them remotely, the result of such advantages is that UAVs can be used in a tremendous number of roles with little or no risk in terms of U.S. casualties.

UAVs can conceivably fill almost every role that is currently accomplished by the U.S. Air Force, including intelligence, surveillance, reconnaissance, communications relay, kinetic operations, and resupply.[82]

Only operations where considerable judgment and adaptation are required, such as search and rescue, are probably beyond the short-term potential for UAVs. These craft, owing to their increased range and loiter time, can also fulfill many of the roles more effectively and efficiently than manned aircraft can. Pilots simply cannot stay awake and stay efficient for missions lasting days, which is the length of some UAVs' missions.[83]

Unfortunately individuals who see a moral or ethical problem with using robots to kill people question the combat role of UCAVs. Whether these individuals are simply against war in general and are using the issue to attempt to limit U.S. power or military power, or whether they have real moralistic issues with autonomous killing, combat roles for UAVs are under scrutiny.[84] The potential variety of ordnance carried and their extremely long loiter times make the close air support role for UCAVs an ideal support platform for U.S. infantry. There is little likelihood that the U.S. military will contemplate any voluntary ban on UCAVs. Political actors in Washington are also unlikely to ban UCAVs given the U.S. military's support of their utility, but their roles and the level of autonomy they are allowed may be limited.[85] It is likely that a person will always have to be in the loop when it comes to killing human beings.[86]

VII. Operationally Offensive, Tactically Defensive Concept and Technology

To use the operationally offensive, tactically defensive concept to its maximum potential requires the development of new technical tools. Today the U.S. military could employ the concept to a certain degree, and arguably it was used in Iraq and is being employed in Afghanistan. However, as pointed out earlier, there are risks associated with the concept's implementation. Without the proper tools, those risks increase dramatically.

The U.S. military has the technical tools to employ the operationally offensive, tactically defensive concept successfully, but the tools are not as advanced as they need to be for the concept's widespread application. The current sensor nets, UAVs, and platoon-level indirect fire weapons would all need to be improved for the concept to be widely employed. The required improvements pose engineering problems rather than conceptual problems. Given enough money and time, which translates into political capital, engineering problems can be solved. Developing the tools to complement conventional needs, or to serve a dual role for counterinsurgency and conventional conflicts, present the real challenge. The new tools would be extremely hard to justify without a dual role.

Implementing the operationally offensive, tactically defensive concept should not initially be based primarily upon technology. The concept has to be grounded in sound strategy to determine if it is appropriate for a given counterinsurgency conflict. Only in a conflict where insurgents are threatening local population centers, rural or urban, *and* the population is willing to tolerate or work with counterinsurgency forces can the concept be employed with any hope of success. As an example, this strategy may not be appropriate for U.S. forces to use in pursuing the Taliban and al-Qaeda in the Pashtun areas of Pakistan, where 90 percent of the population believes it is wrong to cooperate with U.S. forces.[87]

The U.S. force structure would not necessarily have to be radically changed to exercise the operationally offensive, tactically defensive concept. To secure a rural area, for instance, say a force dispatches two squads to each village in question. They should have prior solid intelligence that the villagers are sympathetic to the host nation government. The squads' movement to the village needs to be accomplished as quickly as possible to keep the chances of insurgent ambush or sabotage to a minimum, so they do so by a rotary wing insertion. Following good small unit tactics, the men first establish communications after arriving at the village.[88] Rather than relying on a standard radio check, the team checks the connectivity of a central hub to their sensors, weapons, and intelligence networks and the establishment of a link to a close air support network. Setting up and maintaining these connections should always be the soldiers' first priority because they constitute the team's security. Quite literally the connectivity of their systems is their life. Ideally the main communications hub should be an easily understandable and user-friendly interface as anything too complex runs a risk of causing confusion. The interface also needs to be rugged to ensure longevity and cost-effectiveness and to be usable by an eighteen-year-old soldier of (at least) average intelligence.[89]

The interface must give the operator the ability to receive and process data from all of the deployed sensors. These sensors are directly linked to the interface and available for review by anyone on the network with the appropriate clearance. This type of feed is similar to the one system remote video terminal used in Iraq to give ground troops a direct feed from a UAV operating in their area.[90] The sensors feeding into the hub should be varied in terms of what the troops have to analyze. The sensors' deployment should provide as much overlap as reasonably possible but still be left to the soldiers' discretion as they are in the best position to determine likely avenues of approach by insurgent forces. Many bases in Afghanistan already use a thermal imaging system to scan for approaching insurgents.[91]

The sensors' job is to provide supplemental intelligence that complements the HUMINT that the soldiers receive from the locals. The team's security is based on space and the ability to trade time for that space as an enemy force closes with the team. Sensors themselves are expected to detect enemy movements, classify or measure the phenomena with a high degree of fidelity, and present the information through the user interface in an easy-to-understand format. However, the sensors will never be as good at understanding local trends as going directly to villagers for information. It may seem to be expecting too much of eighteen-year-old soldiers to deploy and monitor high-tech equipment and to integrate the sensor information with local knowledge, but with the appropriate training and the solid leadership of their noncommissioned officers, they should be able to successfully accomplish this part of the mission.

The hub and the soldiers need to solidly integrate into the village. Language-learning software, cultural-learning software, local dress, and food lessons can all be rendered into a digital format for the team to study under the direction of their NCOs. A separate interface also could be available for distance learning, which would be no different from taking a regular online course.[92] A pocket-size language translator, with follow-on technical tools, would be all that is needed to get started and ensure that even an average American soldier is capable of interacting with the locals.[93]

As the team becomes more integrated into the local milieu, the HUMINT available to the team should improve, and their knowledge of the local situation will ensure better technical sensing. Patterns will begin to emerge in terms of where villagers go to tend flocks and chop wood or which days people travel to get goods from other towns or villages. As these patterns become evident, adjusting the sensors can make them more effective. If the acoustic sensors register an anomaly that is classified as a Tajik speaker, the team will know if the individual is a regular visitor or a person of interest who should be watched. If a local shepherd reports seeing men in a certain area, the squad can remotely seed that area with sensors. The technical and the cultural tools thus combine to form a synergy that reinforces each other and provides the team with the best intelligence possible.

As the sensor network around the village grows, the information obtained creates a picture of normal activity. Anomalies and phenomena that are outside the norm becomes readily apparent to experienced team members, or the network automatically flags them itself. To ensure that competency stays high, all the team members cannot leave at the same time to be replaced by another team. As with conventional units, when an entire team rotates out at one time, the knowledge base and social connections that help leverage the sensor technology are lost as the new team attempts to build new relationships and make sense of the sensor data without the benefit of the previous team's cultural learning.[94] Extending tours to two or even three years, even though fifteen-month tours in Iraq caused problems, and ensuring that no more than two members of the team rotate out at any given time should provide continuity to make maximum use of the sensor network and of established social connections.[95]

The most dangerous situation that an embedded team could face, and is indeed the same threat U.S. Army Special Forces teams confront, is one where the village turns on the team.[96] Generally a village turns on a team because a member of the team has done something extraordinarily wrong. Even in such a serious situation, however, technology and HUMINT can mitigate the chance of a catastrophe. Acoustic sensing that listens for key words such as "kill the Americans" would be timely and a possible lifesaver. If the team has built the appropriate social connections, then they should have some warning.

A platoon leader, usually a lieutenant, may be responsible for four villages with his men being divided evenly. The individual networks can be tailored for each village while maintaining interconnectedness. The hub of each village should be remotely accessible by the platoon leader so he can make adjustments to dispositions. The sensor networks of each village should be coordinated to overlap as much as possible for, as with overlapping fields of fire, overlapping sensor images provides a unit with security. Then the platoon leader would be available to move between villages, meet with local leaders, organize militias, attempt to fulfill requests from locals, and generally win hearts and minds.

The entire point of the sensor network is not to kill, even though that would certainly be an ancillary benefit, but to gather information. If villages have up to a ten-mile sensing radius, then the network could possibly pick up insurgent activity that has nothing to do with the village itself. Enemy movement may not require an airstrike, but the information is always valuable. The sensor coverage of the platoon leader's four villages could be more than sixteen hundred square miles.

The number of sensors required is impossible to calculate because a team could always use additional sensors. Meanwhile, those same sensors, if not the same number, are compatible with a conventional mission. The teams in the villages can be recombined into a standard line unit fairly easily with the majority of their sensors put into storage and some kept for the unit's use. Most soldiers who are accustomed to using sensors likely will want continue to use them for increased situational awareness. So in that respect, the investment in research and development of the sensors is dual use.

The ability of a team to create an observed space around itself, especially out to a distance of ten miles, provides more security and combat effectiveness than a larger force could. If insurgents move into the sensing zone, the soldiers would have enough time to evaluate their options. One mile an hour is a very good speed for moving tactically through flat terrain, but in jungles, swamps, and mountainous areas, even that speed may be unattainable without compromising security.[97] Assume a gap occurs in the sensor network, a couple of sensors fail, or insurgents close on a village from an unexpected direction, and the insurgents are not noticed until they were only two miles from the village rather than the expected ten miles. The team still has roughly two hours to put the word out that an attack is forthcoming. The sensors could track the insurgents while a quick reaction force could fly into the village, air weapons teams could be dispatched, or indirect fire brought to bear while the team confirms that the signal is, in fact, an enemy formation. If the insurgents do not retreat by the same route, then they will find themselves in the middle of a sensor network and unlikely to make it the eight miles out alive.

As noted earlier, the importance of the sensor network is not to trap and kill the insurgents but to provide time and space in which to save the village from insurgent influence. Simply using sensor networks to kill insurgents does not bring the operationally offensive, tactically defensive concept into harmony with counterinsurgency strategy. The sensors provide the security for a small number of soldiers through information without overly risking defeat in detail. Owing to the inherent limitations in manpower, the small number of soldiers in each team is necessary to ensure coverage of a large number of rural settlements. It may be a bit too grandiose to refer to strategic sensors for small squad deployment, but it is imperative that the sensors further the strategy and not just the tactics of killing people.

One of the biggest problems that soldiers face in an isolated rural settlement is resupply. In the network example, the soldiers in the village have to ensure that their footprint is as small as possible and their consumption of supplies fairly meager, assuming the operationally offensive, tactically defensive concept is implemented on a large scale instead of incrementally. Batteries for their equipment need to be rechargeable, preferably through solar power.[98] The soldiers should buy much of their food locally to provide the village with an influx of cash, as well as to limit resupply tonnages to a minimum. They should purify water from local sources on site rather than purify it elsewhere and bring it in. The patrol base should be located next to the village or in it if possible. Constructing the base should be done with local materials whenever possible and preferably with materials purchased from villagers. All of these steps are important to keep start-up and resupply tonnages low, to support the local economy, and to build goodwill.

As a practical measure, resupplying outposts should not be done on the ground. As stated, the opportunities for insurgents to attack resupply convoys, along with the logistics of the convoys, make them undesirable. In

applying the operationally offensive, tactically defensive concept to counterinsurgency warfare, the sheer number of villages that require supplies
make road-bound resupply almost impossible. Further combat engineers
have to clear the roads to each far-flung village before the actual resupply
can take place.[99] The only practical way to resupply so many outposts is
to do it by air.

Abandoning roads also has the effect of removing the majority of the
threats from IEDs, which have been favored by insurgents in Iraq and
Afghanistan.[100] If insurgents are not able to attack U.S. military personnel remotely, they will be forced into direct fire engagements against a
technically superior opponent and possibly incur unmanageable casualties.
More important the insurgents will have to close with U.S. forces that
were in a defensive posture and surrounded by sensor networks. In such
a scenario, even attempting to close with the Americans' position is no
guarantee of getting into a firefight. It is quite likely that the insurgents
will be detected and attacked from air assets or indirect fire. Indirect fire
will probably become the insurgents' favored method for attacking U.S.
forces, but such a strategy risks alienating the local people if U.S. forces
are close to or in villages. Indirect attacks are also hazardous as they reveal
the insurgents' position through counter-battery radar and detectable
muzzle flashes.[101] Moving logistical supply away from roads to support
the operationally offensive, tactically defensive concept presumably has
cascading effects for the entire conflict.

Resupply by air runs the risk of exposing the strategy and lines of communication to interdiction from the ground. As the Soviets experienced
in Afghanistan in the 1980s, a strategy based on technical superiority has
the potential to be one technical widget away from a failed strategy. It
is doubtful that insurgents could develop a surface-to-air missile system
capable of shooting down high-flying aircraft; however, a peer competitor
could supply insurgents with a capability such the SA-7, as the Soviets did
during the American conflict in Vietnam.[102] But even a truly advanced
surface to air missile may not be an effective weapon to interdict air-
delivered supplies to remote villages.

By resupplying villages via GPS-guided parafoils, an aircraft can release
its supply pallet from up to nine miles away.[103] The parafoil travels the

distance to the village as it falls. An aircraft flying ten miles away and at an altitude of more than twenty thousand feet in the middle of the night is not a realistic target for interdiction. With each pallet capable of delivering up to twenty-eight thousand pounds of supplies, a single pallet has the potential to deliver twenty-eight hundred pounds for each team member in the village. The supplementary food, sensors, and mail could last a single individual a long time. It logically follows that the difficulty in resupplying dispersed, small garrisons is not the actual delivery of supplies but, due to the large number of outposts, the servicing all of the garrisons. Again providing the airlift required is simply a matter of funding priorities. More important there is no substantial detriment to having a logistically heavy air component as a part of the U.S. Air Force. The same air lift that would facilitate the supply requirements of small garrisons engaged in an operationally offensive, tactically defensive counterinsurgency effort is equally applicable for supporting a conventional war or raiding expeditions.

Much of the work to enhance effective garrison resupply efforts could be automated to reduce manpower requirements in support units, whose personnel all require logistical support themselves. Every soldier in support units requires food, clothing, weapons, equipment, and protection, as well as morale items such as mail. Eliminating the support personnel, when applicable, has a cascading effect of reducing the overall support requirements for a campaign. It is therefore advantageous to reduce to the minimum required personnel who provide effective combat service support to line units or, in this case, to dispersed teams.

The Cobra Brass infrared-sensing satellite is the type of tool that could enhance the streamlining of logistical support for teams in the villages. While the actual capabilities of the system remain classified, it is capable of distinguishing between different thermal events. In the case of incoming indirect fire, IR systems can classify the type of weapon used by the IR signature of the muzzle blast and warn soldiers that incoming fire is imminent within the given range of the specific weapon fired.[104] The warning comes while the incoming rounds are still in the air, thus reducing casualties.[105] The system's potential application in this role could increase the survivability of U.S. forces and the village populations they are supposed to protect. This technology also has a logistical application if the system is enhanced.

The Cobra Brass system can survey an area, look for IR signatures of large thermal phenomena, and classify the event in terms of the artillery type fired. The same technology, if improved, could see smaller IR events in the same area and classify them as small arms fire, grenades, flares, outgoing mortars, or any other IR event.[106] With the system accurately identifying each event during the course of an engagement, the information could be tallied and processed for resupply requirements during the engagement. As known levels of ammunition hit different levels of depletion, the need for resupply can become more critical. The system, or a processing capability of a complementary system that collects and analyzes the data for the Cobra Brass system, could then send out a resupply order with an appropriate level of urgency for the garrison and its supporting units in hostile contact. That order for resupply could be received at a central processing point, where the specific numbers and types of ammunition used could be palletized, depending on the urgency of need, and put on a resupply flight for GPS-guided delivery to the forces in contact.

Similar automated resupply for firebases using indirect fire weapons to support the village garrisons could be a continuous and accurate aspect of IR monitoring. Every round expended by the firebase could be counted and classified for resupply when the known stock of ammunition reaches a certain level of depletion. Ideally the numbers of rounds used, and therefore needing to be replaced, will continue to decline as precision-guided weapons become smaller and more accurate. In short any ammunition with a thermal signature that is fired should be able to be detected and classified for timely and accurate automated resupply.

The automation of logistical support is ideal for reducing manpower and ensuring efficiency of resupply to village teams. Consider the Worldport facility of United Parcel Service (UPS). It sorts, palletizes, and ships up to 416,000 packages of cargo every hour aboard 130 constantly rotating aircraft.[107] All of that cargo is processed and shipped by only 20,000 UPS employees utilizing the company's air fleet of more than 1,000 aircraft.[108] Senders of the cargo have options that translate well into priority classifications of resupply need.[109] If a sender needs the shipment to reach its recipient soon but is in no real rush, the delivery can take days. In another instance, the company can fulfill a sender's request and deliver an important item the next day. A team in a village requesting extra concertina wire often does not need it delivered immediately, but it might not particularly want to wait more than twelve to twenty-four hours before receiving replacement thermal imaging equipment that the team relies on for its security. Certainly the team would not want to wait that long for ammunition in the middle of a firefight. Using the UPS model, logistical support could have several air transport assets always on reserve for dire situations, with the most critically urgent requests being resupplied in hours or less.

In an operationally offensive, tactically defensive counterinsurgency effort, not all resupply would have to take place with air assets. It is not reasonable to expect every piece of hardware and every support item to be moved exclusively by air. Realistically, ground transport only needs to be reconsidered in areas with insurgent activity. If a port is secured and insurgent activity removed from the port city and immediate surrounding areas, the majority of goods can be shipped through the port facilities. As areas are secured farther out, some vehicle traffic could be warranted. Meanwhile, in areas where the operationally offensive, tactically defensive concept is employed or in areas of clear insurgent control, ground transport should be avoided. Such restrictions will likely entail a rethinking of the appropriate amount of kit a soldier needs in the U.S. infantry.

UAVs and UCAVs are two of the most promising technical aspects when employing an operationally offensive, tactically defensive concept to counterinsurgency warfare. Unmanned vehicles have such a wide array of possible applications that their support is invaluable. UAVs could resupply troops by air, allowing for a large fleet of aircraft without a massive number of pilots and aircrews to deliver the required supplies, which the U.S. Marines have already done on a small scale. An automated resupply system could prepackage required pallets to be loaded onto a remotely operated, or preferably autonomous, UAV for the delivery run. UAVs have already demonstrated a capability of taking off, navigating, and landing without any guidance or input from human controllers.[110] Their long flight duration of ninety hours or more would make them ideally suited for the delivery, and perhaps retrieval, of a large number of GPS-guided resupply pallets.[111] The result would be an almost continuous stream of resupply missions with the majority of the work being done by automated equipment. The reduced manpower requirements and risk to aircrews and the potential increase in efficiency will provide strategic and political dividends.

UAVs could also be part of the sensor net that provides the teams in villages with an enhanced ability to trade time for space to protect rural populations and provide increased security. It would likely be beyond the scope of a small team to fly a UAV continuously to watch for insurgent incursions, but a small, portable UAV—similar to the Raven in size but with more range and loiter time—could be a useful supplementary reconnaissance tool for investigating sensor events.[112] It is not reasonable for a two- or three-man patrol to walk ten miles from a village to investigate whether a large insurgent formation is in the area. The time and risk could be prohibitive even though going on foot would always remain an option. It would be far more effective to use the technology of a UAV to investigate the area; to use its multiple spectrums and onboard sensors to detect visual, radar, IR, RF, or magnetic signatures; and to serve as a communications relay to spread the intelligence gained, as the Shadow UAV relayed communications in Iraq.[113] If the UAV found nothing, then it could return to the village while the sensor net continued to monitor the area for other events. If the UAV spotted an enemy formation that was clearly identifiable as hostile, then it could act as a forward air controller to coordinate air weapons teams or to spot for indirect fire missions so as to engage and eliminate the insurgents.[114] Other UAVs not directly under the team's control could be tasked from higher echelons to investigate sensor anomalies noticed by the village team. With their extremely long loiter time, UAVs are far more effective than manned platforms are in a sensor role. UCAVs could serve the same function and with the obvious advantage of being able to engage targets of opportunity or to provide some modicum of close air support.

The ability to put UAVs and UCAVs into dangerous situations has the potential to make medical evacuations easier than they are currently. The pilot of a medical evacuation helicopter has the final say on whether a landing zone is too dangerous for an attempt to evacuate wounded soldiers.[115] While there is tremendous pressure to take a wounded soldier out of harm's way and to medical treatment as quickly as possible, the pilot must consider the welfare of his crew and, more important, how the tactical situation on the ground could be altered if he takes too big a risk and gets shot down. As was seen in Mogadishu in 1993, an entire operation can be suddenly thrown askew when a helicopter is shot down.[116] Saving the life of the wounded soldier is important but not if it results in the aircrew and a dozen other soldiers being killed in an attempt to save a single man. UAVs have the potential to reduce the problem of deciding whether to risk evacuating a wounded soldier.[117] If the soldier is going to die from his wounds unless he receives medical attention, there is no down side to trying to remove him from the battle by UAV; moreover, if the UAV is shot down going in, it does not really matter. If the UAV and the soldier are lost coming out, an airstrike can destroy the site and limit potential propaganda by insurgents who may reach the crash site first. Either way the soldier is dead or gets medical treatment, and other soldiers or airmen do not incur any risk. Every seriously wounded infantryman would intuitively take that risk. With rotary wing UAVs, such as the Fire Scout with a six-hundred-pound lift capacity, it could be possible in extreme circumstances to extract a wounded soldier or villager.[118]

There is a continuing debate as to what degree the United States, as a culture and a military power, wants to trust autonomous machines to operate beyond the supervision of humans. Modern commercial airliners are not flown by people for the majority of their flights.[119] People are not required for an aircraft to take off, fly to its destination, or land. It would be more efficient and cheaper for the airline to operate without humans in the cockpit, but is it realistic to expect people to trust a machine that much?[120] We have all had computers crash, cell phones drop calls, and a tire go flat. Trusting an airplane to travel autonomously between two points is one thing; programming a machine to kill anything it detects in a specified area, even with very narrowly defined parameters, is entirely different. However, there may be no choice as technology could potentially far surpass, in capability, what is feasible by man. If the technology exists to build, maintain, and operate a fleet of 100,000 UAVs to accomplish the myriad of air missions they have the potential to fulfill, can the U.S. military realistically decide not to employ this technology because it does not have 100,000 trained pilots? If the U.S. Army is willing to field autonomous UAVs to fill the same role as the U.S. Air Force UAVs, and in greater numbers, without pilots, how long can the U.S. Air Force resist the technical change?

The future of UAVs is assured even if the roles they fill and the degree of unsupervised action they can take are still unclear. The military utility of the systems, along with the numerous civilian applications, makes the technology too versatile and valuable to eliminate in the interest of "lawfare" or idealism. To contextualize the role of UAVs in an operationally offensive, tactically defensive counterinsurgency effort is similar to trying to predict the future with a certain degree of accuracy; it is foolish and likely to lead to errors. The only sure aspect of UAVs is that, as with the airplane, their roles and uses will vary widely and their capabilities will continue to increase. UAVs, as well as UCAVs, easily fit into a dual role for increased effectiveness for both counterinsurgency and conventional environments.

VIII. Technology and the Future of Counterinsurgency

As appealing as the idea of a revolution in military affairs is, there will be no "before" state and "end" state to the revolution, only a continuing exponential growth in the potential of technology. A clear and concise definition of technologic change, neat in its parameters and foreseeable in scope, would be helpful in making sense of the chaos and dynamic change wrought by technology. Labels, categories, and specific time horizons attempt to bring some of the chaos into focus but, unfortunately, lead to an oversimplification and overestimation of technology in general. Technology is important for the U.S. military and will continue to be so as long as the U.S. military rightfully exploits its lead in military-related technologies.

The degree to which technology is incorporated into U.S. force structure, and the creative applications of that technology, will play an important part in the development of U.S. military capabilities. In turn the military capabilities will directly impact the strategic options available to policy makers in Washington. The course of research and development and the funding priorities, however, will ultimately determine whether technology can be appropriately applied to gain maximum leverage in a counterinsurgency environment. The termination of the Crusader self-propelled artillery system is an example of eliminated funding of an outdated conventional weapon system (outdated in its role rather than in the technology incorporated in the system).[121] The name "Crusader" was almost as egregious as the system's original sixty-two-ton weight, considering it would likely have been used in the Middle East and required far too much logistical tonnage.[122] Eliminating a program, however, is not the same as funding solid counterinsurgency technologies with dual application.

Technologies that further counterinsurgency operations are ones that focus on ISR, precision strike, and those that limit targets of opportunity for insurgents and enhance the capabilities and protection of the individual infantryman. In general nothing in these requirements excludes the systems from being partly or wholly transferable from the counterinsurgency realm to the conventional realm. The amount of logistical airlift may be different, but no military is going to complain that it has too much airlift. The number of sensors that the U.S. infantry requires in a conventional role will not be as high as it needs in a counterinsurgency role, but no infantry unit is going to complain that it has excessive situational awareness. Although too much information can make for bad intelligence, there is no such thing as too much good intelligence.

Technology alone will not make the difference between success and failure in a counterinsurgency effort. No matter how capable the weapon or sensor systems are, the tactics, operations, and strategy must work in harmony to support the policy objectives.

The U.S. military is and will continue to be casualty shy as public support may wane as casualties mount.[123] Nightly news reports will cover any conflict and dutifully keep track of the human cost of the engagement in which the United States is involved. The deaths of service personnel cost money and political capital; more important, they have the potential to make the U.S. military too risk averse to accomplish the policy objectives. Dispersing U.S. troops to rural villages in ten-man teams would have to endure a withering storm of criticism from political pundits and armchair generals. American soldiers would be killed, some villages would be over-run, sensitive technical items lost, and, more important, some U.S. soldiers would be listed as missing in action. The key in applying the operationally offensive, tactically defensive concept would be to start small and make sure that enthusiasm does not exceed the U.S. military's technical grasp. As the military addresses issues and shortcomings, the counterinsurgency environment will become a living laboratory for experimentation and adaptation by U.S. forces.

Focusing on a specific technical widget is not, and will not, be helpful. For instance, although it was not the only piece of technical hardware that could have affected the war's outcome, the Stinger missile that the mujahideen used against the Soviets in Afghanistan altered the war.[124] It did not do so because it had a range of more than eleven thousand feet or because it was a man-portable device; rather, it reversed the course of the war because it changed the tactical, operational, and strategic options available to the Soviet military.[125] The technology forced the conflict to assume an attritional context that did not favor the Soviet forces.

How a technology impacts the harmony of the levels of a war vis-à-vis the policy objectives—not its photogenic or specific technical aspects—is the true measure of the technology's worth. The U.S. military's body armor, for instance, is some of the best in the world. Nylon was used in the flak jackets during the Vietnam conflict.[126] Kevlar replaced the nylon, and later on trauma plates made of ceramics supplemented the standard Kev-lar vests.[127] The future may see Spectra or PBO (polybenzoxazole) fiber hybrids, perhaps incorporating nanotube or carbon fiber technologies to enhance the effectiveness of body armor.[128] None of these fibers has any meaning or validity as a war winner or silver bullet, but body armor reduces casualties, enhances unit cohesion, boosts morale, and enhances combat effectiveness. Historically it has allowed fewer soldiers to accomplish what would once have taken more soldiers to do. The value of body armor is not in the technology that goes into the specific piece of equipment but in the way it allows tactical efforts to be more in harmony with strategic direction and policy objectives.

As the U.S. military grapples with future counterinsurgency roles and demands, it has the opportunity not to focus on the evolutionary and incremental enhancement of conventional legacy systems simply because the systems are a known capability. Technology, especially if it is going to be properly applied to counterinsurgency warfare, must be seen as an enabler of strategy and policy through sound tactical application. Clear policy objectives and coherent strategy must precede the research and development of technology to produce a sensible and applicable military capability. This point does not support Rumsfeld's observation that you go to war with the military you have and not the military you want; in fact, it does the opposite.[129] The U.S. military engages in counterinsurgency operations when directed by policy makers, and it could have *exactly* the military it wants if it is given clear policy and strategic direction on which to base its expenditures and research. Bureaucratic inertia in the military is as detrimental as poor civilian oversight is, but in the end the U.S. military as an organization should be prepared to fight not only the type of conflict it is comfortable with but, more important, any type it confronts. The integration of technology, solid civilian guidance, and a break with rigidity in force structure are required to leverage technology appropriately to apply an operationally offensive, tactically defensive concept to counterinsurgency warfare.

5   Contextual and Other Issues Related to U.S. Counterinsurgency Warfare

> Much recent work in sociology of science, history of technology, anthropology of medicine, cultural studies, feminism, and political philosophy has been a revolt against simplification. The argument has been that the world is complex and it shouldn't be tamed too much—and certainly not to the point where simplification becomes an impediment to understanding
>
> ANNEMARIE MOL

I. Enemy Tactics and Strategy

In this chapter the listing of contextual issues related to implementing an operationally offensive, tactically defensive concept to counterinsurgency strategy is not meant to be exhaustive by any means. It more aptly reflects tangential issues that are pertinent to and illustrative of the complexity and interconnectedness of counterinsurgency strategy. As with any application of strategy, the number of facets and the aspects of each facet that can impact a strategic plan are almost infinite. Just as every tangential issue impacting the concept cannot be exhaustively identified and analyzed, the overall strategy will have nuances and characteristics unique to any counterinsurgency effort in which the United States engages.

Mao's three stages of warfare, as an overall blueprint for how to conduct an insurgency, is a logical and tested method, albeit dated. Ideally the third stage entails the insurgents moving from a stalemate posture to a conventional organization and effort to inflict a decisive defeat on the government forces.[1] This strategy is not as valid today as it was historically. Insurgents are able to coalesce into platoon- and company-size elements, as was seen in the attack on the Wanat outpost in Afghanistan in 2008.[2] However, insurgents are presented with a conundrum when they try to operate in large conventional units. Given that each individual soldier or vehicle represents a phenomenon that may be distinguished by a sensor system, when the numbers of soldiers and vehicles increase, the corresponding number of sensing opportunities also increases. As a general rule, the larger a formation the easier it is to detect with technical means. Likewise, movement is also directly related to sensing opportunities. Increased insurgent movement allows sensors to spot that movement through coherent change detection; and again, the larger the number of moving insurgents the larger the number of sensing opportunities available. To implement Mao's third stage of revolutionary warfare would theoretically require insurgents to form large conventional units to achieve decisive results on the battlefield. In fact, however, this shift would ensure their annihilation as a military force although, as the Tet Offensive demonstrated, military defeats can become political victories.[3]

Small unit actions of company-size elements, such as those that attacked Outpost Wanat in Afghanistan, or even smaller ones are a far cry from a conventional force capable of engaging the U.S. military in decisive conventional campaigns. Insurgent groups without the direct support of a nation-state that has a tremendous production capacity and is adjacent to the area of conflict have no chance of a conventional victory against the U.S. military. Even with favorable conditions, such as the North Vietnamese enjoyed, the insurgent forces in conventional formations were simply no match for the technical advantages of the U.S. military in the 1960s and 1970s.[4] As the Vietnam conflict continued, the North Vietnamese and the Vietcong were able to move in and out of Mao's three stages of warfare when in different areas and when appropriate to local conditions. This flexibility allowed the communist forces to prolong the conflict and respond dynamically to U.S. efforts; however, they never fully realized Mao's third stage to any degree of success. The third stage instead offered opportunities to look strong and score political success through the media during operations such as the Tet Offensive and Khe Sanh.[5]

In Afghanistan Mao's third stage will not be achieved and successfully implemented against the U.S. military, but that stage may be meaningless. The U.S. military never lost an above-company-level engagement in

Vietnam, yet its failure to achieve U.S. policy objectives was absolute.[6] This discrepancy does not mean that Mao's three stages are somehow flawed. Mao would likely be delighted if an adversary could be relied on to abandon its counterinsurgency effort while it was still in the second stage. In short the U.S. military cannot be defeated by a conventional insurgent force in a conventional set-piece battle, but the United States is likely to abandon any effort that lasts too long in the stalemate of Mao's second stage. This point is likely to hold true without regard to the cause of the conflict: pressure built to abandon the counterinsurgency effort in Afghanistan even though it was the result of a direct attack on U.S. soil.[7]

Beyond the issue of a population who cannot be won over to support the host nation government, it is unlikely that the American voting public will tolerate an indefinite and costly counterinsurgency effort.[8] In this regard when confronted with Mao's three stages of revolutionary warfare, the operationally offensive, tactically defensive concept, while being the best way to leverage technology in counterinsurgency operations, will never have a positive effect on public opinion about a counterinsurgency effort. An effort that is stuck in the second stage for any number of reasons may not garner a decisive advantage through applying the operationally offensive, tactically defensive concept. The reasons may include political grievances that originally caused the insurgency are not being addressed, an inability of the U.S. military to convince local people that it is operating in their best interests, or an unwillingness of U.S. policy makers to address foreign sanctuaries where the insurgents have a safe haven. The concept may buy time for a political resolution while also securing a traditionally vulnerable segment of society, but the concept alone will not solve larger strategic issues. Through denying the insurgents contact with the population, the concept is likely to move the conflict back to Mao's first stage of strategic defensiveness when it is linked with other counterinsurgency practices, and that possibility alone should make it a valuable tool.[9]

Since Mao's conventional third stage of revolutionary warfare is not a practical objective for insurgents fighting against the technical superiority of the U.S. military, the real measure of effectiveness is whether an insurgency is in the first or second stage. The U.S. military is doing well if, in conjunction with host nation forces, it can move an insurgency from a stalemate back to a defensive and evasion mode of operations or if it can prevent the insurgency's moving from Mao's first stage to the second. In the first stage, the insurgency is vulnerable and cannot effectively threaten the host nation government while trying to recruit members.[10] This vulnerability offers the host nation time to address issues that may be fueling the insurgency and to train and field its forces to keep the insurgents off balance, as was seen in Iraq after the surge. Employing the operationally offensive, tactically defensive concept not only will hasten the insurgents' move back to the first stage by denying them potential support but also will hamper the insurgency from regrouping in rural areas.

The second stage of Mao's revolutionary warfare may be seen as the new third stage in that it is the decisive stage of the conflict rather than a preparatory phase leading to a later decisive conventional campaign. It would be reasonable for an insurgent group to recognize that Mao's strategy has merit, but any conventional effort is unlikely to succeed as long as American troops are directly involved in the conflict. As the Taliban have likely considered doing in Afghanistan, the insurgents could trade time for casualties knowing that the support of the American people is not limitless.[11] The insurgency faces the risk that the host nation government could effectively institute reforms while training and fielding more capable units, but that risk may be unavoidable. Eventually U.S. forces will have to stop taking casualties or be forced by popular demand to go home. After the departure of U.S. forces, the insurgency could move to the third stage, but by then their victory may be a foregone conclusion.

Typical metrics such as American casualties or U.S. dollars spent on a counterinsurgency effort will not provide any insight as to which of Mao's stages an insurgency is in. As such it will always be hard to contextualize or explain cogently to the American public, much less a twenty-four-hour news media, the progress or status of a counterinsurgency effort. The concept of Mao's three stages has no relevance to the average person's daily life. Therefore, contextualizing the operationally offensive, tactically defensive concept in Mao's stages offers no solution to the inevitable problem of waning American support in a prolonged second stage. Even metrics used by the U.S. military, such as tracking significant acts of violence, do not necessarily offer a good method for tracking progress in a counterinsurgency conflict.[12] Consider President Obama's decision to publicly announce the start of troop withdrawals from Afghanistan starting in 2011. While it is possible that his statement was an effort at strategic deception, it may have caused the Taliban and al-Qaeda to reduce attacks and gather strength for operations after U.S. forces depart. In this context it may appear that the insurgency is losing momentum as violent attacks decrease and casualties decline if the coalition and host nation forces do not actively push into insurgent-controlled areas. This development, therefore, may create a false impression that the conflict is moving toward Mao's first stage. However, without a U.S. presence in rural areas, the strength and support offered by the populace to the Taliban and al-Qaeda cannot be known, much less contextualized for the American public or policy makers.

Arguably some question whether Mao's three stages of revolutionary warfare are still relevant today. If his third stage is no longer a viable choice for insurgents, and a prolonged second stage will not be decisive in any reasonable amount of time, are his three stages really a framework worthy of emulation? The answer should be yes. Not only do the three stages offer a strategic context to better understand the conflict and its progression, but also time is a relative concept for Mao.[13] While American policy makers and the public in general have very short attention spans and want conflicts to be short, efficient, and decisive, it does not mean that somehow warfare must be the same. Provided the insurgency is on a firm foundation, whether a counterinsurgency struggle is in its first, second, or even third decade is immaterial to the insurgent. As long as recruitment is viable, safe havens and supplies are available, and some modicum of popular support is present, the insurgents can afford to feel confident that their political objectives are attainable. So a twenty- or thirty-yearlong stage 2, while not a first choice, does not matter if the motivation to fight is strong enough, as is demonstrated by FARC (the Revolutionary Armed Forces of Colombia).[14] If the insurgents realize that stage 2 must continue as long as U.S. forces are helping the host nation, the insurgency, as a whole, should feel confident that it can outlast the U.S. forces' presence. Whether the insurgents will be able to move to Mao's third stage or will stay in stage 2 after U.S. forces depart is an entirely different matter. The insurgents need worry only if they find themselves moving back to stage 1 with the prospect of the population turning against their cause as happened in Iraq in the summer and autumn of 2010.[15]

Provided that the technical needs outlined in chapter 4 have been met, the operationally offensive, tactically defensive concept is employable in stage 1 of Mao's revolutionary warfare as a preventative measure for vulnerable areas. The concept's application is a necessity in stage 2 as the vulnerability of the rural population and its support potential for the insurgency grows. The concept, however, becomes less desirable the stronger an insurgency becomes. If an insurgency is ready to move into phase 3, or is deeply committed to it, the insurgents likely have substantial resources, manpower, and support at their disposal just as the North Vietnamese had in 1951.[16] The rural villages likely are already linked to the insurgency in some way and therefore probably would not change allegiance to support host nation or coalition forces. Further in stage 3, the insurgency should have the ability to attack multiple targets simultaneously and in strength, resulting in the possibility that some village teams may be overrun as demands for close air support would be very high. Inserting village teams into a high-threat environment is probably asking too much of reorganized line units; instead, these operations stay within the realm of traditional U.S. Special Operations Forces, which are specifically trained to work in such areas.[17]

The operationally offensive, tactically defensive concept is still useful even if an insurgency is moving into or is already in Mao's third stage of revolutionary warfare. The line units that would typically be dispersed as village teams would need to defeat the insurgent formations that coalesced into conventional or quasi-conventional structures. This kind of conventional warfare will lead to insurgent formations being defeated and fragmenting and will be followed by a return to stage 2 as insurgents use guerrilla tactics. The fragmentation will lead to a dispersal of insurgent elements into areas that are unoccupied by counterinsurgency forces, such as rural population centers, in an effort to regroup as al-Qaeda in Iraq did after American troops took control of Baquba in 2008.[18] As the conflict moves back to or into stage 2, the concept should be implemented while the insurgent forces are vulnerable. The final objective of the operationally offensive, tactically defensive concept is, ultimately and preferably quickly, to move the conflict to Mao's stage 1 so that the host nation forces can take the lead in the conflict and provide security while a political solution is found.

II. Countering Enemy Tactics

Kinetics, or kinetic operations, refers to operations that involve shooting or lethal violence and therefore killing.[19] It specifically refers to the kinetic energy delivered by a moving object impacting another body. In typical defense jargon, it sounds scientific and new while benefiting from a slightly more politically correct connotation. At its heart, the word is useful because it alludes to the reality of warfare in that delivering energy into bodies to create wounds has always been a part of warfare. The word is also detrimental in that it attempts to classify a part of warfare, or operations, as being somehow distinct from other aspects. Again it shows a general inability to see war holistically with all the contradictions and gray areas that are typical of U.S. military thinking. Kinetic operations are dangerous, and military personnel often see them as the role of true warriors as opposed to their conducting civil affairs. Journalists also keep the jargon alive. They often seek out kinetic operations to cover as they are likely to get more airtime than reporting other rather mundane aspects of any given conflict. Kinetics, however, are important in counterinsurgency operations because any type of warfare is still warfare with its unchanging character of applying fear and violence (force) to achieve political objectives.

The operationally offensive, tactically defensive concept is not a counterinsurgency strategy. As stated previously, it is a concept or doctrine that allows the leveraging of technical superiority in such a way as to ensure tactical harmony with solid strategy. Whether the concept is employed using small teams in villages in a rural setting or large forces in an urban setting (though both must happen simultaneously for the full benefit to be realized), it is about securing the populace to separate the people from the insurgents according to sound counterinsurgency practice.[20] But with that said, counterinsurgency operations are centered not just on providing social security but also on conducting warfare. It is imperative that insurgent forces be attacked with the aim of killing irreconcilable elements or, at a minimum, of incarcerating them.[21]

Traditionally insurgents have had the benefit of moving about with relative impunity. Insurgents typically are far more familiar with the terrain and social milieu than the counterinsurgency forces that are trying to find them. Often counterinsurgency forces—specifically, the French Union forces' Mobile Groups in the early 1950s in Indochina and the U.S. conventional forces in Vietnam in the 1960s and 1970s—have uselessly scoured the countryside for insurgent formations.[22] If counterinsurgency forces were to make direct contact with a large insurgent formation, having the firepower of an infantry battalion available would be beneficial. However, the slow speed and obvious nature of an infantry battalion almost assure that any insurgents in the area will be able to elude the counterinsurgency forces.[23] The goal of these forces was to make contact with the enemy and decisively beat them by bringing to bear the superior firepower accessible to counterinsurgency forces. In short they were kinetic operations though often without much kinetics.

Blindly flailing about jungles and swamps in fruitless attempts to bring superior firepower to bear is actually a somewhat puzzling exercise. These attempted kinetic operations can only be explained by assuming the practitioners were conventional military thinkers or planners without a grasp of the tactical and/or strategic realities of counterinsurgency operations. In the counterinsurgency realm there is simply no reasonable explanation for such exercises. If the intention is to interdict lines of communications or destroy base areas that store supplies, weapons, and matériel, then using conventional forces does make sense. This attempt was illustrated in Vietnam by the U.S. military and Army of the Republic of Vietnam in the Cambodian campaign and in Indochina by French Union forces in a push toward Thái Nguyên.[24] However, counterinsurgency forces' patrolling in great numbers in the mistaken belief that they will somehow retain the initiative is counterproductive at best and wholly incompetent at worst.

Kinetic operations need to be intelligence driven. Acquiring unreliable and untimely intelligence that insurgent forces are in an area is not grounds for a battalion air assault to scour the countryside for several weeks so as to appear to be doing something. A far better use of the battalion is to secure local population centers and send out small reconnaissance elements to patrol and identify lines of communications used by the enemy forces, thereby providing intelligence for interdiction. If the insurgent forces choose to offer battle for the population centers, then they will be at a disadvantage. The operation is still kinetic in that it is meant to force the insurgent elements to attack but also offers the benefits of the operationally offensive, tactically defensive concept. Generally speaking during the U.S. intervention in Vietnam, the process of shuffling troops around to await an enemy attack would have been seen as apathetic and not doing enough—as would be evident from the lack of a good body count, which was such an important metric at that time.[25] Good actionable intelligence from all available sources is the prerequisite for kinetic operations. Without that intelligence, sitting in population centers, buying the local drink of choice, and making people's lives better or more secure are far better options than doing much else. In short sometimes doing nothing is better than doing something.[26]

One of the problems with kinetic operations in a counterinsurgency effort is the issue of speed. Once actionable intelligence is obtained that a target is in a specific location or that insurgent forces are using a specific location, as opposed to a general area, movement to the location and engagement must be done as quickly as possible. In the U.S. and French efforts in Southeast Asia, the kinetic operations were often compromised from the start, with landing zone watchers alerting local forces when counterinsurgency forces were active.[27] The time it takes for counterinsurgency forces to move to an area and engage enemy forces directly correlates to the time available to the insurgent forces to vacate the area if they choose not to engage in a firefight. Likewise, the larger the counterinsurgency force used, the longer the transport times will be until they are ready to engage or pursue insurgent forces. Ideally the forces used to engage insurgents should be as small as possible, coupled with specific intelligence to allow a raid to happen as quickly and directly to the target as possible, especially if the desired outcome is the capture or killing of a specific insurgent leader.

Technology has the potential to alter the way in which kinetic operations are conducted just as it does with the defensive deployment of village teams in support of the operationally offensive, tactically defensive concept. If specific intelligence lets American commanders know that insurgents are using an area as a base for supplies, training, or rest, there is no longer an overriding need to put large numbers of soldiers on the ground to clear the designated area. Small, superbly equipped and trained U.S. Special Operations Forces can be inserted nearby, can move into the area, and can direct precision-guided munitions to targets in much the same way as the U.S. Special Forces ODA 555 did in the opening phase of the war in Afghanistan in 2001.[28] The detachment was able to identify targets around Kabul for precision-guided munitions, resulting in an estimated twenty-two hundred Taliban and al-Qaeda casualties.[29] The firepower available to a small detachment of soldiers with continuous close air support is more than enough to inflict high numbers of casualties if the enemy's location is actually known.

Reconnaissance in enemy terrain is a highly specialized endeavor that is not suitable for most regular conventional units. Some of the most highly experienced soldiers are devoted solely to reconnaissance.[30] The rigors of such missions combined with the need for stealth require discipline and maturity. While many aspects of a reconnaissance mission will be technically driven, such as air insertion and encrypted burst communications, much of the soldiers' time is spent in quiet physical exertion and control.[31] The payoff of such missions can be intelligence that is otherwise unattainable. There is little substitute for thinking, reasoning individuals who are personally observing events with their own eyes in real time.

The use of U.S. Special Operations Forces in a reconnaissance and kinetic role is not without risks, which are similar to those faced by small conventional teams put into villages as part of the operationally offensive, tactically defensive concept. The units are small, without much indigenous firepower, and are often in areas where extraction could be problematic. If a small special operations team is found, as a SEAL team in Afghanistan was in June 2005, it is very vulnerable to being overrun.[32] The teams generally have to rely on training, stealth, and technology to survive. Small teams that can move unobserved are capable of proceeding to a target area very quickly but are incapable of clearing areas of insurgent fighters. They can organize and direct airstrikes; perhaps assault a small compound, depending on the situation; and provide invaluable intelligence, but they are generally not the equivalent of an air assault infantry unit.[33]

If it is determined that a conventional infantry force is required to clear the area, the only practical way to rapidly assault an enemy insurgent base area in rough terrain is through air mobility.[34] While urban and rural population centers can be defended by technically enhanced conventional soldiers, clearing and holding insurgent base areas usually require conventional units. If a base area is the target, speed may not be as essential as in the case of a time-sensitive target because the insurgents are unlikely to be able to remove their stored supplies quickly. However, with operational security maintained in the planning and execution of a given mission, speedy execution will translate directly into a more effective kinetic operation.[35]

Kinetic operations should always be a tactical component in support of sound counterinsurgency strategy. Killing insurgents is useful but not always required. If sensors near a village or in a pass detect insurgent activity, killing the insurgents may not be the best way to handle the situation, even if the intelligence supplied by the sensors, locals, or special operations teams provides specific actionable information. Allowing the insurgents to continue while tracking them could lead to an understanding of their plans. Letting a supply route be established, used, and perhaps built up before interdicting it could waste insurgents' time and effort. Watching an insurgent visit a village to see if the locals report him or deny his presence could offer valuable insight into their sympathies. Allowing insurgents to operate in an area that village teams cannot cover so that teams can be established in another area unhindered would be more important than trying to kill every insurgent the team found. Kinetic operations do have a strong place in counterinsurgency operations, especially in support of vulnerable village teams employing the operationally offensive, tactically defensive concept, but they should always support strategy while leveraging technology. Arguably the only time a kinetic operation should be a must is to prevent the intimidation of local people who are willing to support the host nation government.

III. Counterinsurgent Manpower

One issue that could have a dramatic effect on the viability of the operationally offensive, tactically defensive concept in counterinsurgency warfare is whether conventional infantry soldiers are capable of surviving and thriving in relative isolation. Using line infantry as village teams in many ways requires that they act as a traditional special forces team with typically little support and interaction with other units. Several questions then arise: Are the soldiers mature enough to handle cultural immersion? How badly would morale suffer with the attendant risks of slipping discipline and substance abuse? Is the typical soldier capable of working with the high-tech equipment that will ensure the security of the village team? Such issues could be a higher concern than technical possibilities when determining if the operationally offensive, tactically defensive concept is viable.

It is not possible to take a historical period and examine it to determine whether soldiers would be capable of implementing the concept for counterinsurgency warfare. Any attempt to do so would be the equivalent of trying to compare apples and oranges. American soldiers today use and are exposed to far more technology than soldiers from any historical period did. The soldiers' awareness of technical capabilities, however, does not mean that they could easily be trained to utilize and maintain high-tech kit. It is fair to ask whether a soldier who is aware of technology is any better prepared to use it. Beyond the issue of a historical technical context, soldiers today are far more knowledgeable of the basics of counterinsurgency doctrine than at any point in U.S. history. The laws governing their behavior are fairly strictly enforced, but that has not always been the case.[36] The awareness that foreign cultures matter is also probably higher than in the past. All of these factors have an impact on the suitability of U.S. troops to implement the concept and to operate as members of village teams.

The only period that has any relevance in determining the suitability of line infantrymen to employ the concept is the current period in which it is being contemplated. As soldiers integrate more advanced systems into the force structure, and the inevitable growth in technical capabilities continues, soldiers likely will become more technically capable and mature. In a future, highly technical, and demanding U.S. military, implementing the concept would likely be easier than it is today, provided that the enabling technologies are developed and fielded. The question has been moot historically as the technical tools had not been developed; therefore, comparing a World War II infantryman with a current infantryman is pointless when determining the viability of line infantry as village team members.

The largest hurdle to implementing the concept is whether the average infantryman is sufficiently mature to operate effectively as a member of a village team. Consider the level of maturity of an average special forces team member at thirty-one years of age in comparison to the average line infantry private at twenty-one.[37] The maturity difference raises the realistic question of whether a nineteen-year-old male is mentally equipped to deal with the rigors of being on a village team. Maturity, coupled with an ability to be calm and collected, is a serious matter, especially after a team member has offended a village population who can turn on the entire team. Every misstep in behavior, even unintentional, must be addressed to the satisfaction of the villagers and their leaders. Flexibility, adaptability, and a willingness to learn are all valuable traits that most nineteen-year-old infantrymen have yet to develop to their fullest capacity.

Soldiers are trainable, and they will meet the expectations that are placed upon them. They may not try terribly hard to exceed those expectations unless a suitable motivating factor, such as their lives being in danger, is present. The standard pace for an infantryman is a twenty-minute mile in combat gear.[38] It is a training standard but becomes a meaningless number when a soldier's life is at risk. In a life-threatening situation, a soldier will move at whatever speed necessary or drop trying. Like standards, requirements can be fluid. It is possible to join the U.S. Army up to the age of forty-two as of the autumn of 2010.[39] If more mature soldiers are needed in a more advanced military, the minimum age requirement can be adjusted upward to fill the ranks with older, more educated, and, more important, mature recruits.

Such drastic measures as possibly hampering recruitment by disqualifying younger and willing potential volunteers may not be needed. Certainly some recruits would not be viable on village teams for any number of reasons; however, even the average line infantryman is capable of learning. Just because a soldier is part of a village team does not mean that the soldier would go without tutoring. As mentioned in chapter 4, distance learning could be a valuable tool to enhance a village team member's effectiveness, but personal mentoring is also a possibility.

Historically the primary job of U.S. Army Special Forces teams has been teaching.[40] Teaching local villagers while attending to their needs is an important role. Likewise, teaching and advising village teams could be an even more effective use of their time. In essence the special forces team members would lend their experience and maturity to help the village team with difficult issues and serve as quality control, ensuring the village teams are doing their jobs to the best of their abilities. A single special forces team could have a large number of villages that its members are responsible for visiting and overseeing and advising the respective village teams. If a situation or reports of problems emerge from a specific village, the special forces team responsible for the area in which the village is located could be dispatched to troubleshoot the situation. Every special forces team has an officer, thus disgruntled local leaders gain the benefit of *wasta*, or "respect" or "influence," by having their issue addressed by someone of a superior rank to anyone on the village team.[41] Even if there is no issue, the sudden and unpredicted arrival of a special forces team, which interacts with the local population and mentors the village team, would demonstrate that the village teams are not truly alone.

Currently women do not serve in combat units in the U.S. Army. While the rule is often ignored when the need for a specific military occupational specialty is desperately required for a particular mission, village teams would be foolish to exclude women.[42] The debate over women in combat is beyond the scope of this book, and the argument put forward here is not an attempt to enter that debate. However, without addressing the possible negative effects of women in combat units in general, the exclusion of women from village teams has two major detrimental effects for the viability of the operationally offensive, tactically defensive concept.

Foremost by excluding women from village teams, the net effect would limit the potential talent pool to fill village team spots. Depending on the makeup of the team, and perhaps location-specific requirements, a female sociologist, anthropologist, or linguist would be far more valuable than a nineteen-year-old rifleman. Keeping in mind that the village team would rely on technology for its survival, the possible upper body physical weakness of a woman would probably be less important than her computer or electronic skills. Removing a large capable segment of people from a talent pool is not necessarily the wisest choice when the mission, according to military doctrine, comes first.

The exclusion of women can also limit the human intelligence available to a village team to 50 percent of the potential. In traditionalist societies such as Iraq and Afghanistan, American male soldiers cannot mingle and interact freely with the segregated females of a village. The private views of local villagers and leaders, revealed at home, are an otherwise unavailable source of valuable intelligence. Incorporating a woman or women into the village team offers the potential to glean that intelligence and form strong personal bonds between the team and the village.

Many male soldiers, however, would not welcome women into the exclusively male arena of combat arms.[43] Likewise, many commanders would rightly worry about the possibility of sexual tension adversely impacting unit cohesion and combat effectiveness.[44] Do these detriments outweigh the advantages of having women included in village teams? The answer to that question is likely to reflect mainstream American cultural attitudes, which are constantly in flux. The real concern of the presence of women affecting unit morale cannot be ignored when considering the isolation and the length of time for which a village team would logically be embedded in the village.

Soldiers' morale is always a concern of commanders. As morale drops there is usually a corresponding drop in combat effectiveness and discipline in a unit. Isolation and a lack of entertainment can be compensated through regimentation and by keeping soldiers as busy as possible.[45] Busy work, however, will not help morale. It is similar to the old adage that beatings will commence until morale improves. Ultimately it would be up to the professionalism of the NCOs and platoon leaders to monitor and maintain the morale of the village teams. The deeper the village team is enmeshed into the daily activities and lives of the villagers, the more likely they will find activities and opportunities that make the time more bearable.

Through remote entertainment and connectivity, technology has the potential to alleviate the boredom that leads to morale problems. A video game console would not only provide entertainment for the soldiers but also allow bonding with village members who would undoubtedly enjoy the novelty of the games as well. A small digital projector could provide the same entertainment for both villagers and team members. Novel technologies such as a digital Koran could also have an enormous impact on the team's relationship with the villagers. Power usage for all items would obviously need to be kept to a minimum.

Technology could also be used to combat possible substance abuse by bored or dispirited troops. Mandatory drug testing with small, portable drug-testing units could potentially identify abuse before it becomes rampant and a serious problem.[46] The negative effects of low morale and drug use by American soldiers in Vietnam should not be allowed to exist in the village team structure. Any means to combat it, technical or otherwise, must be employed, or the security of the village teams would be compromised.

One option to prevent soldiers from feeling that a two- to three-year stint in a far-off village is unfair—feelings that will turn to resentment and lower morale—is to ask for volunteers. Bonus pay, special privileges, and distinctive headgear or badges can all be used to attract the soldiers desired for specific units. As Napoleon said, "Men will fight long and hard for a . . . ribbon."[47] Soldiers who volunteer for a specific duty are far more apt to be content in the execution of the unit's missions. For many soldiers, the unregimented and potentially low-key, away-from-the-flagpole nature of a village team may appeal to them. Those soldiers who are not inclined to such duty could still be effectively used in more traditional roles in a counterinsurgency effort, such as static defense in a larger urban environment, which offers less isolation and a shorter rotation. Even in such roles, they can still contribute to the overall operationally offensive, tactically defensive concept for counterinsurgency.

Soldiers who are not comfortable with or trusting in technology would not be as suited to a village team as they might be to a quick reaction force or mobile strike unit. However, soldiers would not have to be highly skilled to use technology effectively. The onus would be upon engineers to make the technology as user friendly and integrated as possible. Civilian sensor-integration companies have already figured out the importance of simplicity and clarity in sensor displays.[48] By making the displays, or any other technical tool, specifically for use by nontechnically oriented soldiers, their learning curve can be minimized and training time reduced. Making the technology as user friendly as possible has the net effect of expanding the potential talent pool that can be recruited into the village teams. Therefore, the complexity of the technology used should not be a factor in determining a soldier's suitability for inclusion in village teams. Rather, the pressure should be on the engineers to support the soldier's needs.

IV. Insurgent Safe Havens

Any facet of counterinsurgency operations can seem fairly straightforward when examined in isolation; yet all aspects of counterinsurgency operations, as with any form of warfare, are interlinked and should not be considered in isolation. Minute details have the potential to obscure the larger picture. Insurgencies are dynamic in nature, often with multiple groups working for a similar cause. They may rely on each other when the need arises and act independently when it serves their purposes. Above all, insurgencies will always seek support from whatever sources are available. The third parties that insurgencies rely on may play a dominant role in the insurgency, as North Vietnam did in its support of the Vietcong, or a cursory role, as drug traffickers do in supplying financing to the Taliban or as Iranian backers do in supporting some Shia insurgents in Iraq.[49] A coherent counterinsurgency strategy should consider how to deal with third-party interference.

Diplomatic efforts will always be a part of the state's interaction to limit outside support of insurgent forces; however, diplomatic efforts are beyond the scope of this book, as they are not inherently driven by technology. Third-party technical support can have a profound effect on counterinsurgency forces as the technically sophisticated systems that insurgents use typically will come from a state-sponsored third party. In some cases they may be an adaptation of commonly available off-the-shelf technology, such as the use of pagers and cell phones as components of IEDs. The mujahideen fighting against Soviet troops in the 1980s could not have produced an equivalent to the Stinger missile, which was one of the most advanced MANPADs in the world at the time and too advanced for the Soviets to counter.[50] Iranian support of Iraqi Shia militia Jaysh al-Mahdi has dramatically improved its IED effectiveness by providing explosively formed penetrators (EFPs), which the Iraqi insurgents could not manufacture themselves.[51] The North Vietnamese could not have produced the SA-2 surface-to-air missile that the Soviet Union provided to them in the late 1960s and 1970s.[52] These examples are important in that they demonstrate that low-tech insurgencies do not have to stay low tech in every system they use if they have the support of a technically capable third party.

Most troubling for the counterinsurgency forces is that a single technical system can drastically alter the viability of the counterinsurgency effort. As the limitations of politically correct doctrine were outlined in chapter 4, the Stinger missile is a glaring example of a technically oriented military being stymied by a more technically advanced system for which it had no answer. As the U.S. military found out, there is no technical counter to EFPs, which are capable of destroying even the most heavily armored U.S. military vehicles such as the M1 tank and the MRAP.[53] EFPs did not alter the course of the Iraq War to the degree that the Stinger missile changed the Soviet intervention in Afghanistan, but they do demonstrate that high-tech systems are vulnerable to third party–provided systems.

Since insurgencies are expected to adapt to take advantage of counterinsurgent vulnerabilities, the counterinsurgent forces should be prepared to adapt as well. It is not enough for the counterinsurgency forces to estimate that they have an answer to all the low-tech threats from insurgent groups; they must also take third parties into account. Consider the implications of Iran's reverse engineering the Stinger missiles they acquired from the mujahideen during the Soviet intervention in Afghanistan.[54] The U.S. military should be prepared to counter the threat of Stinger missiles, especially given the historical documentation of their effectiveness in Afghanistan. One way to do so would be to mount BAE Systems' tactical aircraft direct infrared countermeasure system, which is capable of burning out the seeker head of missiles like the Stinger.[55] Northrop Grumman makes a similar system for large aircraft.[56] While it has an answer to the threat of MANPADs, the U.S. military should be ready to adapt to other challenges it cannot predict; but, of course, that effort could be very difficult.

Planners should assume that insurgents will counter the implementation of the operationally offensive, tactically defensive concept. They should also assume that insurgents will use anything at their disposal, as well as that of their likely supporters. Insurgents could use information operations to interfere with the sensor nets surrounding the village teams. Propaganda spread through the Internet and designed to undermine the support for the village teams could weaken U.S. resolve, especially if several village teams were overrun. Indirect fire from third-party territory to limit a kinetic response could also be employed. None of these options would likely prove to be decisive, but when combined they could be a cogent insurgent response. What is assured is that insurgents would be forced to respond to the operationally offensive, tactically defensive concept and would probably do so in an unforeseen way with the support of a third party.

Insurgent groups have traditionally used third-party territory to provide sanctuary for rest, logistical staging, training, and recruitment. The Vietminh had base areas in China, the Vietcong used the sanctuary of Cambodia, the mujahideen used Pakistan during the Soviet intervention, the Shia militias in Iraq have used safe havens in Iran, and the Taliban and al-Qaeda have done the same in Pakistan after the U.S. intervention in Afghanistan began in 2001.[57] Such base areas present the U.S. military with difficult problems given civilian diplomatic concerns. With civilian oversight of the military at all times, diplomatic and political considerations heavily outweigh military considerations when the question of attacking these sanctuaries arises. Widening the Iraq War to include military operations in Iran would have irreversibly altered the course of U.S. counterinsurgency efforts in Iraq. Likewise, military incursions into Cambodia and Laos were difficult political issues in Washington during the U.S. war in Vietnam.[58]

Third-party safe havens are not an insurmountable problem for a counterinsurgency force. They do make kinetic progress against insurgents much more difficult and, if left unaddressed, have the potential to cause long-term problems. However, the goal of a counterinsurgency strategy is to secure the population to allow political solutions, not necessarily to kill insurgents as a first priority. Since the populations who need to be secured are not in the third-party safe havens, it is strategically more important to focus on securing the local people first. Ultimately if the local people are secured, the third-party safe haven is of less importance in the long run. If the local population centers (urban or rural) cannot be secured, then third-party safe havens have the potential to make the counterinsurgency effort futile.

As indicated in the introduction, the United States is committed to operating within and in support of the international system of states. Justifying an intervention in a single state can be difficult, as demonstrated by former secretary of state Colin Powell's briefing to the United Nations Security Council prior to the invasion of Iraq in 2003. Arguably the same effort would then be required to intervene in an adjoining sovereign state if a coalition were to be maintained. The U.S. government quite likely would feel that the political costs of expanding a counterinsurgency effort into an adjacent sovereign state would outweigh the benefits of actually doing so. In such an event, it would be in the U.S. military's best interest to tailor the counterinsurgency strategy to emphasize speed and technology.

As important, technology can offer the U.S. military the potential to be able to ignore third-party safe havens in a short-term time frame. The safe havens are only a detriment if the flow of people and matériel from them cross the border unobserved into the area of conflict. There also could be a large number of general areas to which insurgents could fall back for rest and resupply. If the areas are known and the transit routes observed, however, then the safe havens simply offer help in targeting insurgents by concentrating them in observable areas. As they recross the border into areas designated for engagement, kinetic operations can be used to keep them from closing on population centers.

In whatever manner the U.S. military and civilian policy makers choose to deal with a third party offering insurgents safe havens, good counter-insurgency practice should not be compromised in the given conflict. The safe havens do not fundamentally challenge the viability of the operationally offensive, tactically defensive concept. Safe havens in third-party territory will offer long-term strategic challenges but not tactical challenges if technology has been properly leveraged to support the village teams.

Third-party safe havens do present a problem if they become training grounds for an unlimited number of potential insurgents. During the Soviet intervention in Afghanistan, for example, large numbers of insurgents from all over the Arab world went to Pakistan to participate in the jihad.[59] While technology can offer the U.S. military the ability to employ the operationally offensive, tactically defensive concept to secure population centers, it cannot sustain the Americans' will to participate in a counterinsurgency conflict. Third-party safe havens that can accommodate an unlimited manpower resource, driven by identity or ideology, will eventually overwhelm the Americans' will to continue. As the Americans' will erodes and the prospect of an eventual U.S. withdrawal becomes a reality, the ability of U.S. troops to maintain the local people's support will diminish as well.

The problem for the U.S. military in conducting a counterinsurgency effort against an unlimited supply of manpower in a third-party safe haven is that, short of military intervention, the United States is incapable of forcing compliance from a third-party state. The Soviet Union also faced this dilemma regarding Pakistan's support of the mujahideen. Without the Soviet leadership's political determination to intervene in Pakistan, the Soviets could not force Pakistan to discontinue its support for the mujahideen. With unlimited manpower at Pakistan's disposal and areas to train and equip the fighters beyond the Soviets' reach, the Soviets faced an indefinite conflict. Even with technology properly leveraged, the American people also are unlikely to support an indefinite intervention and, more important, an ever-increasing expansion of the conflict into neighboring sovereign states.

The best way for the U.S. military to handle a third-party safe haven with unlimited manpower available is to devote its resources toward an operationally offensive, tactically defensive concept while exerting the maximum possible pressure for political reform and political solutions by the host nation. The more widespread the implementation of the operationally offensive, tactically defensive concept, the faster security provision and the return of good governance can be. Speed, in this case, has two important motivators: the political will of the American public is exhaustible, and the longer a third-party safe haven is used to train and support insurgents, the larger the challenge will be for the host nation's security forces to cope with the long-term insurgency.

V. Keeping Coalition Partners

The United States will always want to intervene in a conflict as a senior partner in a coalition effort. Even if the coalition is simply a facade, a coalition offers the United States a level of legitimacy when it deploys military forces in pursuit of U.S. national interests.[60] However, coalitions cost U.S. policy makers' strategic leeway as partner concerns have the potential to hamper or constrain military effectiveness due to both political considerations and technical limitations. No coalition partner can afford to ignore domestic political concerns indefinitely as all politics are local. At some point coalition partners, when engaged in a domestically unpopular conflict, will have to question their continued involvement in a U.S. military intervention as the French and Germans did with their involvement in Afghanistan. As such the U.S. military and the political decision makers need to consider carefully whether coalition partners are worth the political cost, can contribute to the overall effort, and bring any detrimental aspects with their involvement. Even if their participation in a coalition is deemed beneficial, the technical question of interoperability would be extremely difficult if the U.S. military were to employ the operationally offensive, tactically defensive concept in a counterinsurgency environment.

Building a coalition has the potential to undermine the ability of the United States to act in defense of its vital national interests. Coalition partners may demand that diplomatic efforts be tried first, or they may push the United States to seek the support of international bodies for a decision to use force prior to the partners' committing to their own participation, as the United Kingdom did before the 2003 invasion of Iraq.[61] In such scenarios the United States is at least tacitly acknowledging that the international body has some authority to approve or disapprove the intended military action. Whether this involvement has a positive or negative effect on U.S. military power is inconsequential, other than that it may directly impact the number of foreign forces and capabilities that could potentially enhance the prospect of success in a given conflict. However, if the United States must surrender the ability to act independently, then almost no support offered by coalition partners is likely to be worth the diminution of freedom of action.

Supposing that coalition partners are found in reasonable numbers and the political price paid for their help does not outweigh the appearance of legitimacy they offer, serious consideration of what they can offer needs to be assessed. Of the forty-two members of the "coalition of the willing" that took part in the invasion of Iraq in 2003, almost all were incapable of providing meaningful combat effectiveness.[62] Some countries, such as France, were specifically seen as unhelpful and not courted for the coalition even though they may have provided more legitimacy.[63] At some point beyond expedient diplomatic considerations, coalitions need to bring some degree of usefulness to military operations, or they are simply window dressing.

On the one hand, counterinsurgency operations with properly leveraged technology are extremely difficult and dangerous in areas contested by insurgent formations. On the other hand, they are relatively easy and safe in areas firmly under the control of the host nation's government. It is still dangerous, but the threat to counterinsurgency forces is much more limited. The use of coalition partners in these safer areas, where the likelihood of combat is low, or the likely need for combat skills and experience is not high—such as the disposition of German troops in northern Afghanistan and French forces around Kabul—does make sense militarily.[64] But certainly a potential for problems exists if combat does occur. French troops made very basic mistakes that resulted in the loss of ten French soldiers in 2008.[65] Having coalition partners occupy the safer areas, thus allowing more U.S. troops to be shifted to areas under contention, would support sound counterinsurgency practice. If the safe areas become contested and coalition partners have severe interoperability problems with U.S. forces, then the coalition quite likely would be a detrimental aspect to the counterinsurgency effort.

It is currently difficult for U.S. forces effectively and efficiently to operate jointly with foreign militaries, with primary exceptions being the UK and Australian forces. A foreign mechanized force without a blue force tracking system in its vehicles can become a detriment rather than an asset as positioning errors and night operations can quickly lead to friendly fire mistakes. French attempts to participate in close air support missions in Operation Anaconda were utterly ineffectual and potentially dangerous due to the lack of precision-guided munitions.[66] In an age of warfare where information and communications are to be decisive tools, a lack of interoperability makes coalition partners less useful.

Coalition participation in the technically leveraged operationally offensive, tactically defensive concept for counterinsurgency warfare is likely to be extremely limited if partners have no interoperability potential. There is some potential for peripheral roles for coalition partners, such as host nation force training or other civil affairs and engineering efforts. However, without interoperability partners can contribute little to population security in areas where the population is actually threatened, to kinetic operations, and even to logistics and combat service support. It is even likely that coalition partners would not want to contribute.

The U.S. intervention in Afghanistan has shown the unwillingness of some countries to engage in combat missions or to sustain casualties. The overwhelming majority of combat fatalities, due to operations in contested areas, have been sustained by the United States and the United Kingdom.[67] The majority of the German people do not support the war in Afghanistan, making casualties politically difficult to sustain.[68] Both the French and the Germans understood clearly that their participation in Afghanistan was in their own security interests as the war was part of a NATO effort. France and Germany want to participate as little as possible, however, and garner continued security benefits through NATO without risking severe domestic outrage.

The dangers and hardships of life for the village teams likely would be impossible for coalition partners to sell politically to their domestic audiences. Tours in the villages would be too long. British soldiers typically had only six-month rotations in Iraq and Afghanistan.[69] The dangers of being overrun or having villages turn on the village teams could be mitigated, but they would occasionally happen. The French would be incapable of sustaining major losses and would be particularly unable to do it under U.S. command without creating division and drama well beyond their usefulness. The sensor technology employed by the village teams and the networking of all pertinent data and communications systems could not feasibly be expected from coalition partners. Without coalition interoperability, even the automated supply system of UAV delivery vehicles and GPS-guided parafoils would likely need less participation from coalition partners to be more effective.

Coalition partners could also present a security risk in an operationally offensive, tactically defensive effort. The research and development of the sensor systems, networking, and communications would represent a substantial U.S. investment. While history shows that the technology will eventually be disseminated, inviting any willing country to participate and potentially acquire the technologies would not be in the best interest of the United States.[70] Theoretically all coalition partners would be allies with similar interests, or they would have little reason to participate. Yet technologies are withheld from some NATO members and more so from random potential participants in a coalition.[71] The degree to which the U.S. military is willing, or can afford, to compromise the technologies that will allow the operationally offensive, tactically defensive concept to be implemented will directly impact the viability of coalition partners to contribute substantially to a counterinsurgency effort.

The United States has several challenges, internationally and diplomatically, to leverage technology for counterinsurgency application. NATO, representing some of the most technically advanced militaries in the world, cannot effectively employ military force specifically to further U.S. national interests. As a collective security organization, its primary purpose is defensive in nature.[72] While this role does not preclude NATO from engaging in counterinsurgency operations, such as in Afghanistan, without popular European support, it will not be able to bring its technical advantages to bear over a long period of sustained effort. To ask that a new NATO standard of interoperability with U.S. forces be achieved through increased research and development for application to long-term counterinsurgency warfare is simply not realistic. Without the technically oriented militaries of Europe being able to assist in applying the operationally offensive, tactically defensive concept in counterinsurgency, there are no practical options other than a unilateral effort by the U.S. military.

NATO also has no apparent gain from participating in counterinsurgency operations other than to maintain a semblance of an alliance through its nominal support of U.S. goals. NATO countries would likely not intervene in a country without the United States if the operation were to lead into a long-term counterinsurgency conflict.[73] Force projection and regional stability outside Europe are not priorities for Europe as a whole. Given the casualty-shy nature of European populations and their political representatives, serious doubts about an effective role for NATO in a high-tech and especially a high-risk counterinsurgency conflict should be kept in mind when building a coalition.

Clearly every NATO member is different regarding its individual willingness and capability to contribute in a counterinsurgency setting. To ignore the differences between the cultures and militaries of the United Kingdom and France, for example, is to make too broad a generalization about NATO members. The willingness of the Tony Blair government to support the invasion of Iraq in 2003 is a clear example of a (perhaps fading) special relationship between the United Kingdom and the United States. That level of cooperation and connection between the two countries is obviously far closer than that between most NATO members and the United States. However, it does demonstrate the need for the United States to think about coalition building, if a coalition is indeed beneficial, in terms of individual sovereign nations rather than of organizations or supranational groups as a whole.

Unilateral action by the U.S. government to engage in hostilities must always remain at the discretion of the U.S. government alone. In general the United States views organizations of plurality and consensus as positive forces; however, the UN does not necessarily enjoy the trust of a large segment of the U.S. population.[74] Whether real or imagined, most Americans view the UN as not having supremacy over the United States and as potentially trying to limit U.S. sovereignty. While Blair was adamant that the United Kingdom needed the UN's approval to invade Iraq in 2003, some individuals in the Bush administration felt that going to the UN, and the opinion of the UN, was useless and immaterial.[75] As interpreted by an organization and its member states, which may have a vested interest in limiting U.S. hard power, international law represents a direct threat to the ability of the United States to defend itself and its vital national interests.

Limiting U.S. hard power, either through insisting that it has no right to wage war or through restricting new technologies used in a conflict, has a direct impact on the viability of coalition building and the operationally offensive, tactically defensive concept in counterinsurgency operations. If, based solely on an arbitrary decision by a few individuals in the United Nations, other nations refuse to enter into conflicts that the United States feels absolutely compelled to enter, coalition building becomes unrealizable. Former secretary-general Kofi Annan voiced the opinion that the 2003 invasion of Iraq was "illegal" because the UN Security Council did not explicitly authorize it.[76] If a future UN secretary-general were to express the same view before the start of a conflict, it may preclude the prospect of building a coalition. Likewise, if a UN secretary-general were to deem illegal certain technologies that the operationally offensive, tactically defensive concept would employ, such as autonomous UCAVs and other robotic systems, the decision could affect the viability of a coalition when it became clear that the U.S. military would use them. Land mines, depleted uranium shells, lasers, UCAVs, cluster munitions, psychotropic drugs, and incendiaries are all military tools that are deemed potentially illegal, or that should be banned, under international law depending on how they are employed.[77] They also happen to be systems for which the U.S. military has a legitimate use. If populations demand that their governments seek UN authorization before joining a U.S.-led coalition, especially in a counterinsurgency conflict, and the UN describes the intentions and tools of the U.S. military as illegal, it may be better for the U.S. government to abandon the idea of legitimacy through coalitions.

VI. Countering U.S. Tactics

If the U.S. military properly leveraged technology for counterinsurgency warfare through the appropriate application of resources, insurgents would find themselves in a very difficult position. Insurgents moving through empty wilderness terrain, always out of contact with local people, are as ineffective as counterinsurgency forces randomly wandering about in search of insurgents. The insurgents must make contact with local people to influence their opinions in favor of the insurgency and to gain support. Ideally by applying the operationally offensive, tactically defensive concept, the insurgents' only viable targets are the village teams and the static defense forces in urban settings. As explained earlier, direct assaults on those soldiers are not only potentially catastrophic but, more important, unlikely to provide contact with the local populations who need to be co-opted into the insurgent cause. With aerial resupply, the viability and utility of IED attacks—the insurgents' most favored method of attack in Iraq and Afghanistan—are dramatically reduced as the only traffic will be that of civilians, whom the insurgents do not want to kill.[78] The farther the insurgents travel from their safe havens, the more likely they are to be detected and intercepted. In short, the apparent options of the insurgents will be extremely limited.

Insurgents are masterful at doing the unexpected and finding unlikely solutions to problems. For example, it is quite likely that military planners instituting an operationally offensive, tactically defensive effort to secure rural villages in a given area would double-check and triple-check their careful plans. They would pore over requirements for logistics, close air support, sensor networks, communications, and civil affairs until they concluded that they could secure the area without much problem. They would likely foresee no insurmountable challenge in insulating the population from the insurgents. They may even come to the conclusion that the insurgents have no options available to them. Then things would start going wrong as the insurgents figured out a way to adapt to the situation. A village would be overrun and a team either lost or put at extreme peril until a quick reaction force could arrive. Lessons learned and the failure of the sensor net would have to be analyzed and remedied. The after-action review, which is a fundamental tool for improving unit performance, would have to be examined for clues as to the insurgent's behavior and then changes made throughout the village teams.[79]

This scenario would be a constant process as the insurgency adapted to the counterinsurgency forces and vice versa. Some of the insurgents' tactics could probably be predicted through war gaming and simple logic; however, due to the social context, many of the adaptations may be impossible to predict. What to an American civilian or a U.S. soldier would seem likely and plausible may be total anathema to a Pakistani villager. Americans may be thinking of how they would try to defeat the sensor nets as that move would be paramount to closing with the village team unobserved. Although a Pakistani may come to the same conclusion, he might think first about who he must engage to convince the villagers that collaborating with the Americans is wrong. The individual may have enormous influence in the local culture and yet may be completely unknown to American forces; therefore, it would be impossible for U.S. forces to predict that individual's involvement. In essence outsiders will never be able to accurately predict the minds, intentions, and problem solving of a foreign people.

Insurgents may be uneducated and rigid in their thoughts, but it does not make them stupid. The Taliban might shun Western-originated technology, but that does not mean they are unable to employ it effectively. Their experience with the Stinger missiles in the 1980s should make that point clear. Their use of the Internet to communicate propaganda is also adept.[80] Al-Qaeda's desire to acquire chemical, biological, and nuclear/radiological weapons is well known although the organization itself is incapable of producing them.[81] When a crafty and motivated insurgent combines with a third-party supplier of technology, the foregone conclusion should be that the insurgents will devise novel and unexpected ways of exploiting the counterinsurgency's weaknesses. These weaknesses could be in theater, where they could deal directly with fielded counterinsurgent forces, or even back in the United States, where they could exploit a possible weakness in the Americans' political will.

It is quite likely that the insurgents will direct some effort at aspects of the conflict that cannot be solved by U.S. military technology. The Qatar-based television network Al Jazeera offers an unfiltered outlet for al-Qaeda propaganda and messages.[82] The U.S. government complained and requested that the station not air messages by al-Qaeda and even struck the network's offices in Kabul in 2001.[83] Al Jazeera refused and left the United States no technical solution to the problem. Of course the United States could have jammed the satellite signal, but the network would have simply used a different Arab news outlet.[84] Deliberately jamming civilian news channels would cause more harm than good for U.S. counterinsurgency efforts and is a losing proposition in the long run. In addition, there is no technical solution to the recruitment activities in the up to twenty thousand madrasas operating in Pakistan, where students are taught primarily the Koran or jihadi ideology to the exclusion of more earthly disciplines such as math, science, and critical thinking.[85] Using technology to target these institutions would be tantamount to bombing children in schools. The U.S. government has pressured the Pakistani government, which has indicated it recognizes this educational system as a problem, to confront these institutions, but the U.S. government and the U.S. military can do little or nothing about it even with current or future technical capabilities.[86]

Fundamentally insurgents fight for political objectives, but most likely those political objectives are the results of ideas and ideologies. No technical system is capable of eradicating an idea or an ideology; therefore, the U.S. military, and especially policy makers, should be very careful about the stated aims of counterinsurgency efforts.[87] Wars on tactics, fundamentalist thought, or concepts in general are not helpful. Such language has a tendency to obscure the purpose of counterinsurgency operations—that is, to buy time for reform and political compromise. A war on terrorism is misleading by its name alone and has done nothing directly related to the security of population centers to further Afghan governance. More important, the strengths of the technically oriented U.S. military cannot be used effectively.

Insurgents' use of propaganda will continue to be a problem, and most probably the United States will not develop a coherent response to it.[88] As insurgents work toward their political aims, they realize that information, or information that supports a specific point of view or ideology, is a more useful tool than bullets are in convincing local people to support the insurgency. More important, the insurgents understand and are able to exploit the cultural perspectives and issues of the local people. Naturally the host nation has that benefit as well, but the insurgency would not exist if it did not have culturally relevant grievances with the host nation government. The grievances may not seem important to the population as a whole, but it must resonate with a fair percentage of the people to take root beyond a terrorism campaign. In one study, half of the insurgencies sampled found less than 10 percent of urban populations supported the insurgencies.[89]

An insurgency would also not be possible if the host nation government had taken reasonable steps to acknowledge the political and ideological concerns that give the insurgency its impetus. Not every ideological position, however, can be accommodated. For example, any reasonable government probably could not oblige a hyper-strict version of Islam. The inability of insurgents to make political accommodations with contrary ideologies does not change the fact that the ideological view of the insurgents may have widespread appeal among some of the population. The insurgent propaganda, therefore, will be culturally relevant and have some drawing power, and the U.S. military is unlikely to be able to counter the situation with the same level of relevance.

Global communications and the propaganda that insurgents disseminate will have a detrimental effect on the ability of the United States to build coalitions. In some respects the insurgents' use of propaganda through global communications will predate the counterinsurgency operations themselves. Most future conflicts likely will develop out of problems that are already being examined and circulated via the medium of global communications. Any U.S. military involvement in a primarily Muslim country will be affected by contemporary propaganda and not just the propaganda of tomorrow. In some respects global communications are allowing the insurgents to shape the human terrain of future counterinsurgency conflicts now, and that human terrain also includes the populations of potential coalition partners in any future U.S. military endeavor.

Consider the reactions of different cultural groups after the attacks of September 11, 2001. Many people in the United States were appalled and horrified that American civilians were targeted in such a manner, yet some voiced the view that the United States was itself to blame because of its policies in the Middle East. Some Europeans expressed their opinion that the United States got what it deserved, conveying a harsher reaction than would have been uttered, even rarely, in the United States. In the Middle East, some people actually danced in the streets with joy after the attack. The driving factor behind the cultural gap was extremely disconnected views of historical events, as well as propaganda. Palestinians would not have danced if they had not already been exposed to propaganda and indoctrination that made thousands of American deaths seem justified.[90]

If propaganda had made some Palestinians feel exultant at the news of the 9/11 attack, other populations in the Middle East surely had the same reaction. So by September 12, 2001, U.S. policy makers may have asked themselves what was the best way to sell the impending U.S. invasion, but by then they were already far behind a propaganda curve that they did not even know existed. Culturally relevant propaganda had been widely disseminated through global communications, and any late response by the United States would have seemed irrelevant, biased, and disingenuous in its timing.

Propaganda need not be true to be effective because a large number of people actively want to believe it. For example, almost 23 percent of Germans think the U.S. government was responsible for the 9/11 attacks.[91] Insane and convoluted conspiracy theories from self-proclaimed experts resonate with some people because they already have a preconceived notion that the U.S. government is shady and untrustworthy. Critical thinking, research, and logic are not a part of the knee-jerk reaction that drives the conspiracy mind-set. Critical thinking is also not a valued strong point in the Arab mentality, owing to the restrictive societal influences in many Arab countries.[92] When individuals with preconceived notions are repeatedly exposed to propaganda that supports their own viewpoints, the propaganda need not contain any more truth than their original erroneous beliefs, and the speed with which the hype is disseminated is likely to make it even less accurate.[93] Therefore, an insurgent response—against which the technical capabilities of the U.S. military are useless—has been, is, and will continue to be believed in the Muslim world. It shapes the human terrain through its globally promulgated propaganda. Worldwide communications could be considered as a continuous strategic delivery system for insurgent propaganda, where the truth of what is delivered is utterly meaningless.

Insurgents understand that perception is reality even if it has no connection with actual facts or events. Anything that resembles the preconceived notions of the audience is at least possible. Globally disseminated propaganda reports that Israel indiscriminately kills civilians. To the mind of a neutral observer this assertion may be true or it may be completely fabricated. The United States is portrayed as supporting

Israel and providing the weapons systems that allow Israel to maintain an unjust military advantage. To an individual who is sympathetic to the Palestinians, these messages may seem reasonable approximations of fact even though they are propaganda and ignore some important points. Having spent decades exposed to images, videos, newspaper articles, and other propaganda showing U.S.-manufactured weapons systems killing Palestinian children and hearing about continuous U.S. impediments to UN resolutions condemning Israel, Arabs have preconceived notions that the United States is unjust. Therefore, insurgents will globally disseminate information about any military operation such the second battle for Fallujah as having been a massacre, and many Arabs will believe it was a massacre. Facts, such as the military's dropping leaflets on the city urging civilians to leave, are not only left out but, more important, are considered irrelevant. It is not the massacre but rather the perception of a massacre targeted at a specific audience in the region, as well as at the global sympathizers, that is important.

Combating propaganda is essential in counterinsurgency operations, but it cannot be done in a politically neutral or politically correct context. Video footage highlighting technical capabilities may seem reassuring to the American people and military professionals. Consider, however, how the gun camera footage of an AH-64 Apache helicopter killing insurgents is contextualized to people already predisposed to seeing the U.S. military as a brutal tool of an unjust occupation.[94] Those individuals do not see bad guys dying; rather, they see devout good guys being massacred. This example is a classic misapplication of technology in counterinsurgency warfare. As pointed out before, cameras are not an inherently unsuitable technology to counterinsurgency operations, nor are the Internet and other global communication networks that disseminate that same footage. The problem is that the technology is not being applied to support sound counterinsurgency practice for strategic gain. The release of such videos, intentional or otherwise, provides the insurgency with propaganda, and the availability of global communications and the lack of sound counterinsurgency practice do the insurgents' work for them.

Unfortunately Western tendencies to voluntarily self-edit images and stories to fit comfortably within politically correct standards are another misapplication of technology. Repeatedly showing pictures of children roasted in ovens, people with hands amputated for smoking, dead and mangled children from insurgent-laid IEDs, mosques damaged by insurgent attacks, girls scarred by acid thrown in their faces for going to school, and similar atrocities committed regularly by insurgents and beaming them throughout the world on the Internet and on television are the correct applications of technology.[95] Responding to insurgents' propaganda by graphically exposing their methods and atrocities denies the insurgents a moral foundation from which they can counter effectively. American news and governmental squeamishness about the realities of warfare do a disservice to strategy. Somehow in the American cultural context, war should be a clean, antiseptic version of a TV series. Airing nightly productions, adding drama to an already dramatic situation by serious reporters depicting clean frames of dangerous undertakings, is not reality. The reporting is only striking enough to hold watchers' attention until the next commercial break and yet devoid of any graphic material that may upset the revenue-providing sponsors and an isolated American public. In maintaining the comfort of a viewing audience, it is almost as though the domestic information battlefield has been deemed a no-go area.

The operationally offensive, tactically defensive counterinsurgency concept requires technology for security, but it also requires strategic support. Inevitable and unpredictable insurgent responses will require flexible counter efforts where they occur—that is, on site—and beyond the geographic boundaries of the conflict. Information technology far removed from the village teams will have a direct impact on everything that relates to them. How the U.S. military and the government react to a dynamic, responsive insurgency will likely have a far greater impact on the village teams than will the issues that they can actually deal with or control.

VII. Maintaining National Will

The political struggles between Democrats and Republicans, or conservatives and liberals, can have unfortunate repercussions on a counterinsurgency effort. With a four-year election cycle, numerous administrations are elected in the time it takes to complete the average fourteen-year counterinsurgency effort. One administration may not favor the political goals of a previous administration. Barack Obama actively campaigned on the promise of withdrawing U.S. troops from Iraq. John McCain, his opponent, generally favored an open-ended timetable that would allow conditions in Iraq to determine the time and pace of U.S. troop withdrawals. For the U.S. military, such opposing views, and the military's requirement to obey the commander in chief to the letter, make devising a coherent strategic plan almost impossible. If the U.S. military plans for a fourteen-year effort in a realistic appraisal of the counterinsurgency task ahead, it simply cannot be expected to change its strategy every four years to meet the latest political whims of Washington and remain effective.

It is as disingenuous as it is simplistic to categorize Republicans as hawks and Democrats as doves, but it often happens. Most terms are descriptive labels for political consumption, yet there is a kernel of truth in most stereotypes. The parties, to some extent, reflect an ideological divide in the United States, with more Republicans than Democrats support-ing the war in Afghanistan in general.[96] Enough American citizens can swing in favor of or against a war effort over the course of fourteen years, however, for the U.S. military to have serious reservations about engaging in counterinsurgency operations.[97] Democrats will call for the end of a war to pick up votes from their more liberal base, as well as from those centrist voters who may be tired of the nightly news reports of casualties and frequent media-inflated situations that appear dire. Unfortunately this type of political pandering, or risk aversion to keep the political cost of casualties low, can have a negative effect on solid counterinsurgency prac-tice. Republicans often say the Democrats are soft, yet more drone attacks from U.S. UAVs operating in Pakistani territory have occurred under the Obama administration than took place under its Republican predeces-sor. Equally disingenuous, Republicans are often depicted as redneck simpletons who are too dumb and overbearing to make rational decisions, much less conduct operations as delicate as counterinsurgency warfare.[98] Yet it was the Bush administration that finally instituted a semblance of coherent counterinsurgency strategy in Iraq under General Petraeus.

Quite likely many members of Congress do not have a grasp of the realities of counterinsurgency warfare. Congress authorized George W. Bush to use military force to remove Saddam Hussein from power in 2002.

After a few years of albeit extremely subpar counterinsurgency practices, many of the same members of Congress, such as Representative John Murtha, were ready to call the effort a failure and blame mismanagement by their political rivals.[99] Without an understanding that counterinsurgency operations are long and tedious and often experience setbacks and unanticipated problems, can members of Congress really be expected to fund the research and development, as well as the fielding, of the technologies required? If members of the U.S. military were to testify before Congress that they were unenthusiastic about the survival prospects of the village teams or about the wisdom of the strategy, could members of Congress resist political gain through interference? If they were told about village teams in an operationally offensive, tactically defensive concept, without an appreciation of technology or the strategy of counterinsurgency conflicts, they would probably scratch their heads in puzzlement and then be disinclined to acquiesce to the appropriations request.

Expecting professional politicians to set aside personal gain when matters as serious as war are on the table is quite likely naive. Expecting reporters who do not understand strategy and are not interested in reporting reality to assess and analyze the war is probably more of the same. However, the people, or the ones who vote, are the final arbiters and decision makers on whether a war can continue. Unfortunately the complexities of counterinsurgency warfare do not fit into the standard eight-second sound bite offered by American media, nor are they explained by the kinetic scenes packaged for prime-time television coverage.[100] The end result is the domestic information battlefield is fought over for partisan political gain rather than engaged as a facet of strategy to leverage against the insurgents. In an ironic twist, as policy analysts in the West debate the merits of dividing the Taliban through an Afghan political process, it would be equally valid for the Taliban to wonder if they could use an information campaign to divide liberals and conservatives in the West. Coalition members would like to reduce the Taliban's strength and ability to continue the conflict, and the Taliban could use dissension and discord to reduce the coalition members' will to continue the conflict.

The U.S. military has not taken casualties at a rate to make the war in Afghanistan indefinitely unsustainable. A total of 1,172 U.S. military personnel died in Afghanistan in the first nine years of Operation Endur-ing Freedom.[101] The number of fatalities is less than the 1,227 U.S. military personnel discharged for homosexuality in 2001 alone.[102] It cannot be argued that the war in Afghanistan is a drain that is decimating the ranks of the U.S. military at a rate of unsustainable losses when the U.S. military voluntarily removed more service members for a sexual preference than it sustained as casualties in nine years. In 2005, the United States spent only slightly more than 4 percent of its GDP on military spending, rank-ing twenty-fifth in the world as a percentage of GDP allocated to military spending; Greece is number 24; and Oman is first at 11.4 percent.[103] At those levels of military expenditure the United States would not bank-rupt itself fighting in Afghanistan. Al-Qaeda is known to operate in the border region between Afghanistan and Pakistan and is estimated to be a serious continuing threat.[104] Given these facts, why did President Obama determine that the U.S. military should start to withdraw U.S. troops from Afghanistan by 2011?[105] Perhaps the reason was not politically motivated toward appeasing anti-war sentiments prior to presidential elections in 2012. Another explanation, however, seemed unlikely.

The political jockeying by conservatives and liberals may be more than a desire for personal gain. It may coincide with a genuine fickleness of the American people. After September 11, 2001, the average citizen was willing to support an administration that was determined and able to engage the enemy wherever found. Tolerance for casualties may have been high and a willingness to sacrifice present. Yet as time passes and the sharpness of the memory fades, the anger and desire for revenge is also likely to fade. A generation is growing up that does not remember the events of Sep-tember 11, 2001, or was not even born at the time. Soon the youngsters will be of age to join the military and to vote for politicians endorsing any number of plans for the war on terrorism. A real question should be considered as to how committed they will be to fight for a cause with which they do not identify. The continued viability of a war is contingent upon popular support, and the individualistic nature of American culture, and society, may make a long counterinsurgency effort anathema to the national character. Certainly the professionalism of the U.S. military and its backbone of NCOs will enable the U.S. military to fight a low-level counterinsurgency conflict for a very long time and likely far longer than the people will allow their leaders to continue it.

Domestic support for a long-term counterinsurgency conflict is also contextual. After the 9/11 attacks, President Bush enjoyed a 90 percent approval rating as he headed into the initial invasion of Afghanistan. With an approval rating that high, almost any military action was feasible. The longevity of the military action was not explicitly specified, nor could it be accurately predicted. The U.S. military could reasonably see its early success in Afghanistan as confirmation that the capabilities and systems used at that time were adequate for responding to the attacks in New York and Washington DC. As time passed, there was no resolution to the war in Afghanistan, and hostilities in Iraq began to sap the support of the American people. While the 9/11 attacks were horrific and had a galvanizing effect for a military response, over time the shock and desire for action faded.

Islamic fundamentalists' killing several thousand people and demolishing a couple of symbolic landmarks came actually as a relief to some analysts in the United States who had closely followed the rhetoric of al-Qaeda. The toll from a well-planned and executed attack employing biological or nuclear weapons would have had been on an entirely different scale. Several hundred thousand dead American citizens as the result of an anthrax attack would provide an entirely different context and provoke a potentially different military response. Certainly the level of revenge demanded by the American populace and the willingness of the American people to engage in a lengthy war would change. Funding would become nearly unlimited, and politically correct doctrine could be ignored with little domestic opposition. Even an event in the United States similar to the 2004 terrorist attack on School Number One in Beslan, Ossetia, which left more than three hundred dead and hundreds wounded, would have a different context from the September 11 attacks.[106] The specific and deliberate attack on innocent children would potentially have the American people's support for a long, sustained counterinsurgency effort. Anger could dictate that the response resemble more of a raid rather than a long-term struggle toward nation building, but political support would likely be absolute with little or no partisan interference.

A war effort or counterinsurgency conflict that was predetermined to be long, with substantial financial investment required and high expectations of casualties, would be understandably unattractive to potential coalition partners. Domestic support, depending on the catalyst for the conflict, may be swayed by a lack of international support and legitimacy. Although domestic support for some level of military response would not require a coalition in order to commence, without coalition partners there could be some hesitation when considering the level of long-term commitment the American people would feel comfortable with while knowing the full burden of the effort would fall on the United States alone.

This chapter demonstrates the difficulty for the United States to participate in counterinsurgency operations with the current force structure. For the United States to be more successful, the reforms indicated here and in chapter 3 would enhance its prospects.

Implications

What has this helpless people done, that they should be driven from their homes, to wander strangers and outcasts and exiles, and to subsist on charity?

MAYOR OF ATLANTA

War is cruelty and you cannot refine it, and those who brought war into our country deserve all the curses and maledictions a people can pour out.

GENERAL SHERMAN

I. Writers and Policy Makers

Many of the writers about small wars and counterinsurgency have made good points but are dated. Robert Taber, Robert Thompson, and David Galula all had contributions to make toward recognizing that the practice of counterinsurgency operations required a different mind-set and emphasis vis-à-vis traditional conventional warfare. The wars of national liberation fomented by communist and anticolonial ideology were a natural process, and they rightly contextualized past counterinsurgency theory. However, the historical and political contexts of national wars of liberation are not constants that will forever define the scope or limits of insurgency and counterinsurgency operations. A people struggling to throw off perceived colonial oppression have a different level of motivation, and their cause has a wider appeal than an extreme reactionary and violent religious fundamentalist movement would. The core idea that the Clausewitzian center of gravity is in the minds of the local people remains a constant in both cases, but the narrative and requirements to succeed can be drastically different.

The optimism of some writers, such as Galula and Thompson, may also be dated. Populations comprising, by and large, very radical individuals whose ideology is based on deeply ingrained identities with religious connotations may never be susceptible to a counterinsurgency effort based on winning the hearts and minds of the local populace. Other writers, such as Taber, are too pessimistic in their appraisal of the futility of counterinsurgency warfare. Not every effort is a lost cause, and, more important, technologies that none of the writers could foresee have the potential to continuously redefine what is possible.

The implications of continuous change in technology and the continuous contextual change of counterinsurgency warfare demand that knowledgeable civilian and military leaders consider strategy rationally as part of a counterinsurgency effort. It is not acceptable for future conventional operations to slowly shift into a counterinsurgency conflict as if by accident and without preplanning, as happened after the 2003 invasion of Iraq.[1] To have a U.S. military surprised at the emergence of an insurgency, coupled with a lack of sound strategy for a counterinsurgency effort, is damning of both the military and the civilian leadership. Future interventions must be better planned and not rely on hope and enthusiasm to mask a lack of technique; otherwise, the American people will rightly hold responsible their elected officials, who will in turn hold military professionals accountable.

The nature of war and of strategy is unchanging. The use of force, as defined by the application, or threat, of fear and violence will always be part and parcel of warfare. Warfare without those two components will cease to be war in whatever form of competition or confrontation it takes. Without fear and violence the primal motivational and coercive nature of war ceases to have any relevance. As moral as liberal or legalistic desires are to eliminate warfare, they are profoundly misguided. War serves political interests and is engaged in to achieve political objectives. Political objectives will never be separated from the social animals that are human beings. Occasionally the motivations to achieve the desired, or even required, political objectives will cause fear and violence to be used as tools. This outlook may be sad, disappointing, and depressing, but it will always remain true.

Strategy too is always available to manage the proper application of force, although it may sometimes not be employed in warfare, to the detriment of whichever party does not use it. The better art of strategy is applied, the more likely an entity is to achieve its political objectives in war. This fairly obvious statement applies to counterinsurgency warfare as much as it does to conventional warfare.

If civilian and military leaders do not recognize the constants of war and strategy, then they will be less able to tailor counterinsurgency practice appropriately and to leverage technology to the context. Public acknowledgement of the inevitability of war is not a politically correct view in the liberal democratic West. Unfortunately a lack of politically viable historical methods of counterinsurgency warfare may also accompany a political leadership's genuine lack of awareness of strategic and military matters. As such, the U.S. military will always be at the mercy of fortune as to whether it will have the required competency in civilian and military leadership to allow it to perform counterinsurgency operations with the appropriate tools.

Leaders must recognize history and use it as a guide rather than a prescription. Assumptions made from a cursory review of historical events will always be examined through the personal filter of the reader. The failure of the United States in holding back communism from South Vietnam does not mean counterinsurgency warfare is a wasted and useless endeavor. Just because technology did not provide a pivotal advantage in Vietnam does not mean that technology cannot be leveraged in counterinsurgency warfare. Conversely, success in the counterinsurgency in Iraq will not automatically mean the same nuances in strategy will work in Afghanistan or even in Iraq twenty years from now.

Appreciation of historical trends must accompany strategic understanding to avoid erroneous assumptions about a potential counterinsurgency effort. Leaders who do not grasp the nuances of a particular situation or conflict will be tempted to review historical data to determine the best courses of action. This assessment alone is better than nothing; however, in doing so, it is possible that leaders will reach a conclusion and decide on a course of action that is not contextually relevant. Historical analysis is required when contemplating or implementing a counterinsurgency strategy, but it must be done effectively through melding the historical information to the many facets of strategy. Otherwise, serious flaws in analysis will result in equally serious flaws in understanding the conflict and lead to assumptions like a correlation between Iraq and Vietnam. Just because apples and pears are fruit does not mean they are anything alike.

The coupling of historically relevant information with strategic understanding and an acceptance of the nature of war should be embraced by both civilian and military leadership. It is probably too much to expect a professional politician to be fully versed in and comfortable with formulating strategy. It is not too much to expect military professionals and competent advisers to help the president and other decision makers to recognize the realities and relevant aspects of a situation prior to implementing a counterinsurgency strategy. A higher caliber of strategic thinking overall is needed in both the civilian and military leadership of the United States than has recently been evident.

II. Proper Counterinsurgency Practice and the U.S. Military

The operationally offensive, tactically defensive concept in counterinsurgency warfare is specifically designed to leverage technology for strategic effect. Technology developed in isolation of a larger need may be reactive to a specific requirement but may lack flexibility. Better engines that are more fuel efficient, radios that are more secure, and canteens that fold and are flexible are all good ideas. They all serve a purpose in each of their narrow niches and help the military to be more effective; however, in isolation the improvements they represent are equivocal. Even if the U.S. military had the most fuel-efficient engines, the radio "after next," or an "objective" canteen, a pertinent question is, so what? The technology needs to support the strategy, even if in doing so it is only enhancing the tactical level of war.

The operationally offensive, tactically defensive concept offers an operational and tactical framework for appropriate technical application. Only through tactical enhancement can an improved strategic practice of counterinsurgency be achieved. Securing pockets of urban populations in a sea of rural insurgent territory is a recipe for failure. The strategy of counterinsurgency is untenable if a large, potentially sympathetic rural population and territory are ignored. In this situation, the tactical execution must support the strategic vision. The only practical way to do so is through a massive number of troops or a technically enhanced smaller number of soldiers. A massive number of troops recruited or drafted into the U.S. military will not be politically acceptable for the foreseeable future, leaving technology as the only reasonable option for increased capability.[2]

By recognizing the necessity of the operationally offensive, tactically defensive concept in counterinsurgency warfare, the U.S. military needs to undergo a dramatic shift in funding, training, and organizational deployment. It is also pertinent to recognize that the current level of risk aversion is counterproductive in counterinsurgency operations. If the current trend continues, political leaders who remain casualty shy will force risk-averse decisions on military leaders that will undermine proper counterinsurgency practice. The number of casualties sustained in Iraq and Afghanistan are extremely low when historically compared to other conflicts. To obtain the political objectives, the primary focus should be to accomplish the mission and not to minimize the casualties.

The validity of the operationally offensive, tactically defensive concept has undergone a historical progression. As the concept is directly linked to technology and the asymmetric technical abilities of adversaries, the number of soldiers required to implement the concept has progressively shrunk with the continued technical growth of the U.S. military. What would have taken a corps to implement 150 years ago required no more than a company of infantry in 1968, in part owing to a generous amount of close air support. The historical trend is not guaranteed to continue at the same rate, and it is not possible to determine specific time horizons at which the concept will mature in different ways. However, the trend is clear, and ignoring the potential in an era of expected counterinsurgency operations is unwise.

Currently the viability of the concept rests in the political will of U.S. policy makers. To a certain degree it is reasonable to expect the U.S. military to implement the concept with the current forces and technology available. Trial efforts to determine specifically needed systems upgrades and to develop tactical and operational requirements are well within the scope of the U.S. military today. Combat trials in the living laboratory of Afghanistan, Somalia, or Mali provide a proving ground second to none. Small numbers of village teams with sensor support could be used effectively to understand future requirements and steer future trials and technology development.

The historical trend indicates that the U.S. military will become more capable of implementing the concept in the future. The U.S. military has continually implemented new technologies to provide new capabilities, although perhaps not always as efficiently as could be desired, and it is forecast to continue on the same trajectory.[3] Potential adversaries, meanwhile, have not always followed the same trends. The technologies used by insurgents today do not greatly differ from the technologies that were available to the Vietcong in the 1960s. With these divergent trends, the asymmetrical advantage in technical capabilities enjoyed by the U.S. military is likely to continue to grow and enhance the possibility of effectively employing the concept.

The U.S. military is uniquely suited to take advantage of the potential of the operationally offensive, tactically defensive concept for counterinsurgency warfare. U.S. military culture embraces technology because of its decisive nature in conventional conflicts. It is also accepted partially as a result of the level of education in the U.S. military and partially as a function of the defense industry's tremendous technical research and development capability in support of the U.S. military. There is no reason to think that the U.S. military will change its perspective on technology any time soon; therefore, it will benefit from continued technical advances.

The U.S. military will always want the most educated and highly motivated soldiers it can find to fill its ranks, but need will always come before desire.[4] It has the advantage of an extremely large and well-educated first world population from which to recruit. In essence its recruitment criteria can be adjusted to acquire specific levels of education and/or experience to best fulfill the given missions. Whether the military is willing to trade off numbers for quality, or increase personnel spending to the levels required to obtain large numbers of highly qualified individuals, is entirely up to budgetary and appropriation considerations; but the potential is latent in the population. If the U.S. military were to implement the changes needed to take full advantage of the operationally offensive, tactically defensive concept, radically new standards and selection criteria would be required.

The defense industry in the United States will continue to offer the U.S. military innovation in whatever direction it requires. The defense industry is a powerful economic force, representing hundreds of billions of dollars being pumped into the U.S. economy. Civilian political needs to give direct support to constituencies through federally funded defense spending ensure continued technically oriented support. The implication is that the U.S. military has nothing to lose by demanding the necessary technical products to implement the counterinsurgency concept as the military will likely receive the political support to obtain what is needed.

III. Lessons from Iraq

Iraq demonstrated that the U.S. military is capable of using line units to conduct sound counterinsurgency practice. Any assumption that line units are unsuitable for counterinsurgency operations is ill founded, because it supposes that line soldiers are incapable of understanding why the rules of engagement are tailored to accomplish counterinsurgency objectives; for practitioners of counterinsurgency who are directly impacted, the reasons and results are tangible. With the solid professional standards of a volunteer force, as opposed to a drafted military, counterinsurgency is not an impossible task for regular line units. As it is very unlikely that the draft will be reimplemented, the U.S. military will continue to be able to rely on enlisted men to act professionally and carry out counterinsurgency operations.

The primary reason the soldiers can be counted on to act in accordance with sound counterinsurgency practice is the professionalism of the NCOs and the control they exert on the troops. As the backbone of the U.S. military, NCOs are the soldiers who enforce standards of conduct. While not autonomous, NCOs are allowed to exercise enough initiative to ensure standards remain high among the troops under their command. By combining professional, career NCOs with a volunteer force of enlisted men, the morale problems that plagued the U.S. military in Vietnam, along with the attendant negative behaviors that damage counterinsurgency operations, can be avoided.

The U.S. military is the most combat-experienced professional force in the world today. Having accrued years of combat experience since 2001 and solid counterinsurgency experience since 2006, the NCOs are in the best position to propagate a legacy of counterinsurgency awareness and practice. New recruits will be quickly exposed to counterinsurgency lessons with the opportunity then to implement them in theater. The level of experience in both the officer and NCO corps of the U.S. military makes implementing the concept on a trial basis feasible and offers high-quality combat experience to help shape the technical requirements.

The counterinsurgency effort in Iraq has shown that the U.S. military is capable of flexibility and adaptability even in mid-conflict. The expectations of a quick conventional victory in Iraq followed by a fairly painless transition of authority to a new and peaceful Iraqi government were naive at best. Accepting the problem, and the need for a counterinsurgency effort, came far too slowly and reluctantly to both the U.S. military and its civilian leadership. It clearly showed that assumptions, coupled with wishful thinking, are no basis for sound strategy. Slow adaptation and a refusal, in the interests of political spin, to recognize the inception of an insurgency only compounded the failure.

Surprisingly the U.S. military was able to adapt to the insurgency and began to implement sound counterinsurgency practice. This flexibility lends further weight to the assertion that regular U.S. line soldiers are capable of counterinsurgency operations. It is also another indictment of much of the U.S. military leadership, which clearly failed to implement the correct strategy prior to General Petraeus's command. The flexibility and adaptability that the U.S. military showed mid-stride in Iraq should be incorporated into the U.S. military as a whole. The likelihood of a strategy having to be adapted, perhaps drastically, should not be resisted or seen as a sign of failure; rather, it should be expected to happen and accepted as reassuring that methods are being refined to handle the specific challenges of a given conflict as they are all unique.

As counterinsurgency operations are planned in the future, a properly prepared U.S. military will be ready for adaptation, and the introduction or large-scale use of the operationally offensive, tactically defensive concept will not be difficult to conceptualize or implement. Funding for its technical requirements will not be seen as useless or misappropriated, and the soldiers who have to implement it will likely be capable of doing so. If the given conflict presents problems that the concept does not fully address, then modifications can be made to tailor it appropriately.

Iraq demonstrated the increased effectiveness of technology for counterinsurgency warfare. Across the entire force structure, technology led to heightened lethality, which in its own way provides security. Precision munitions for fire support also led to the reduced likelihood of civilian casualties, which directly threatens a counterinsurgency effort. UAVs and sensing capabilities have contributed to effective intelligence gathering and provide more precise targeting. The combat effectiveness of U.S. forces has likely had a direct influence on local leaders to engage with U.S. forces constructively rather than kinetically. The technologies used have started to provide the level of connectivity and precision firepower that will allow the operationally offensive, tactically defensive concept to become refined.

Due to the continued conflict in Afghanistan and the other hostilities likely to occur elsewhere, the legacy of Iraq will not result in a compartmentalization and eventual amnesia regarding the counterinsurgency lesson learned by the U.S. military. Likewise, the technologies and techniques developed and refined in Iraq will continue to undergo revision and examination. The technologies and follow-on capabilities will be linked to counterinsurgency doctrine. The forced awareness and on-the-job training that the U.S. military had to painfully undergo in Iraq will have far-reaching and long-lasting effects on the U.S. force structure for many years to come. If the U.S. military demands technical tools for counterinsurgency operations, then the defense industry will oblige them. The industry will build upon the systems already identified in command mission needs statements, thereby giving counterinsurgency technology a life of its own with congressional support.

With the tremendous amount of enlisted and NCO experience in the U.S. military, the legacy of technology from Iraq should not be the exclusive province of generals and civilian strategists. The possibilities for applying technologies should be driven as low as possible—that is, down to corporals and possibly even privates. Technologies that do not have solid applications in the field, as determined by experienced practitioners, should be reexamined. Those systems determined to have merit should be funded and improved. The legacy of Iraq has the possibility to be the first time that serious, large-scale technical capability specifically designed for counterinsurgency could be the norm for the U.S. military. As the future threat of fundamentalist jihadist ideology is likely to remain, such an investment would serve the U.S. military well.[5]

The U.S. military needs to continue fostering increased emphasis on counterinsurgency practice and the awareness of its proper application by line units. The complexity of counterinsurgency operations with an emphasis on civil affairs and social interactions is well outside the U.S. military's comfort zone. As each counterinsurgency conflict is unique, there is no easily implemented training regimen to prepare troops. Troops instead need to be exposed to theory and case studies, which can then be used as guides to critical thought. Such scholastic pursuits for soldiers are possible, if not necessarily commonplace.

Both NCOs and officers in the U.S. military are expected to have pursued further education or have a postsecondary education. The expectation of continued learning by leaders is complementary to the continued propagation and dissemination of counterinsurgency principles. Although counterinsurgency warfare does not need to become a deeply developed discipline in the NCO and enlisted ranks, some continued awareness should be fostered. Counterinsurgency warfare is not rocket science and should not be treated as such. Simple curricula or training programs are more than capable of reinforcing the basics of counterinsurgency.

The military school system can be used as an excellent tool for ensuring the lessons learned from Iraq are not lost to enlisted men. An extra week could be added to military occupation specialties training for combat arms units to deal specifically with counterinsurgency requirements. A separate course for teaching NCOs counterinsurgency principles, and how to teach those principles to enlisted men, could be supplemental to the basic and advanced NCO courses. The implications are that counterinsurgency lessons gained in Iraq can and should be incorporated in the U.S. military educational and training system to ensure longevity.

IV. Limitations of Counterinsurgency Strategy

Counterinsurgency strategy based on the work of such writers as Galula is not the only effective way to conduct a counterinsurgency campaign, but it happens to be the most politically correct and the most likely method to provide long-term stability. The need to engage in such long-term counterinsurgency practices requires a tremendous level of commitment from political leaders and support from the American voting public. The financial investment alone will always have political opponents and a segment of the public questioning the wisdom of the intervention. Therefore, depending on the level of U.S. commitment to norms and standards, as well as the impetus for the counterinsurgency effort, cheaper and faster methods may one day become a possibility.

When the U.S. military is confronted with a population who is unwilling to work with foreign military forces or who views as illegitimate a host nation government that is supported by foreign powers, the application of counterinsurgency practice is problematic.[6] Due to the constraints it must observe, the U.S. military cannot force a recalcitrant civilian population to accept a political outcome through the application of fear and violence. The most troubling aspect of this limitation is that it runs directly counter to the nature of war itself. Ignoring the fact that warfare is about imposing a political will on an adversary runs the risk of engaging in occupational intervention without the possibility of closure through a political settlement.

Obviously the first choice should always be to apply counterinsurgency strategy to build a stable and self-regulating government that conforms to the general rules that govern state behavior in the international system of states. However, if it is not possible to organize and foster such a political entity because the local population refuses to participate for identity-based reasons not open to modification, then the U.S. military has no options left. Civilian political leaders will likely not have the courage to demand that the U.S. military accomplish the political objectives at all costs, leaving the conflict to drag on without resolution.

In a hypothetical situation of a failure in Afghanistan and the possibility of the Taliban, or Afghanistan-based al-Qaeda, acquiring and using biological agents, what possible responses are available to the U.S. military? If counterinsurgency failed to work the first time as a result of Pashtun stubbornness, is there any reason to suppose it would work a second time? If the Taliban effectively sheltered al-Qaeda once and the U.S. military was unable to be successful after thirteen or more years of intervention, would the Taliban have any reason to suddenly deny al-Qaeda sanctuary? Clearly the answer would be no on both accounts. In such a hypothetical situation, it is unimaginable that the United States would sit idly by as biological weapons were manufactured, exported from al-Qaeda safe havens, and used on Western targets around the world. Yet what options would be left when using fear and violence against those who support and give shelter to al-Qaeda and the Taliban is out of the question due to politically correct and legalistic constraints? Scorched-earth practices undeniably would entail huge political costs for the United States internationally, but they would likely be far less onerous than a potentially indefinite series of biological attacks.

The strategic void the U.S. military will face when confronting an uncompromising local population needs to be addressed prior to the United States actually finding itself years into a counterinsurgency conflict. This prerequisite implies the need for debate and the examination of a potentially caustic topic by civilian and military professionals who run the risk of politically correct–driven stigmatization. Just as the U.S. military failed to honestly appraise the likelihood of an insurgency in Iraq after the 2003 invasion, it should not, through optimism, make a similar mistake about counterinsurgency practice.[7]

It is important to note that abandoning counterinsurgency strategy cannot be done on a piecemeal basis or partially. Scorched-earth practices and punitive action applied with blanket coverage will doom a classic counterinsurgency effort. A shift into, or the commencement of, punitive deterrent efforts should happen swiftly and totally, implying a premeditated strategy. The preparation of an intentional, preplanned, and thoroughly examined scorched-earth strategy would be required, but given the political atmosphere and moral self-conception of Western liberal democratic nations in general and the United States in particular, this course of action would also be close to impossible.

Punitive raiding is probably a far more likely application of fear and violence than is a scorched-earth counterinsurgency. The long-term political involvement in a foreign country can be ignored, the attendant financial costs are reduced, the length of the intervention can be unilaterally determined, and the level of destruction predetermined. Its political objective would not be based on the institution of liberal ideals and Western forms of democratically elected governments. The application of fear and violence would clearly indicate that there is a price to be paid for attacks on the

United States and its interests. It would be a return to the true nature of war without an attendant liberal ideology.

Technology has the potential to make raiding strategies far more effective and appealing than a counterinsurgency strategy is. Just as technology has the potential to enhance counterinsurgency strategy through the viability of the operationally offensive, tactically defensive concept, it also can allow punitive raiding without the traditional risk of large ground forces. Small teams directing air-delivered ordnance and air mobility utilizing speed and firepower both reduce the financial and political costs of punitive raiding.

Scorched-earth and raiding strategies are abhorrent to the moral foundations that underpin the professionalism and sense of purpose of the U.S. military. Implementing such strategies should always be well thought out and considered a last resort as it will fundamentally challenge some of the defining aspects of the American character. However, in the absence of another viable option, and coupled with an overriding need to act upon a known threat, there may not be an alternative. More important, it implies that the merits and applicability of the strategies must be examined prior to their actual implementation. To do otherwise would run the risk of an overwhelming backlash both domestically and internationally as propaganda and associated images disseminate through global media.

Without a reexamination of and changes to the rules of armed conflict and international treaty obligations, such as the Geneva Conventions, the U.S. military will not be able to use force to accomplish potential political objectives. The changes cannot be effectively contemplated, debated, and enacted in mid-conflict. Beyond the appearance of self-serving alterations, the process will likely take time and consultation with both domestic political entities and international military partners. Without military and civilian thinkers willing to take the professional risks to push for reexamination, the United States will eventually find itself in a paradoxical situation.

V. Technology

Technology can change the way the United States conducts counterinsurgency operations. The technical capabilities available today have finally started to solve the problems of small, dispersed units being overrun or defeated in detail. The tactical enhancement to provide wide-area coverage of rural populations to maintain separation between insurgents and local populations is revolutionary. As noted earlier, applying the operationally offensive, tactically defensive concept in counterinsurgency warfare may not be appropriate to every cultural group, but the technologies available are maturing to a potentially new level of effectiveness.

For convenience the new possibilities and technical innovations can be labeled; however, the classification of the new technologies and capabilities as a revolution in military affairs should not be taken too literally. There will be no end to the technical innovations that drive military capabilities. It may be easier to comprehend and to sell the need for increased technical capabilities by using readily recognizable and definable labels, but the reality has been, and will continue to be, an exponential increase in technical possibilities. Using the "revolution in military affairs" label may be generally harmless unless it is taken to imply that after the revolution is over, technical innovation can or will cease to compound.

In the sense that the label provides a new emphasis on technical capabilities, which in turn has the potential to increase funding for the research and development of promising technology, it is a benefit. The advantage to the war fighter is in the evolutionary improvement to existing systems and new systems that allow new capabilities. Domestically areas with defense contractors profit from from an influx of federal funds. Highly technical skills are developed, and the demand for high education levels in a workforce reward educational advancement. Technologies developed, such as the global positioning system satellite constellation, have spin-off effects as military capabilities further economic potentials.

The revolution in military affairs label has the negative potential of overemphasizing technology at the expense of strategy. To say that the revolution in military affairs will transform warfare implies that technology is the most important factor in warfare. This emphasis has the potential to further alienate the already sorely neglected role of strategy. Civilian and military leaders who are not versed in strategy will likely make the assumption that the U.S. military, with all its technical capabilities, should be able to accomplish any given mission. Technology, as an enabler of capability, is supplemental to strategy, but strategy should never be compromised by a fascination with technology.

The implication of holding that technology furthers strategy is that independent budgetary requests from the different services can be counterproductive to a strategic end. The U.S. Air Force may desperately want expanded or continued funding for legacy platforms such as the F-22 Raptor and will, therefore, make the aircraft a funding priority.[8] The aircraft is superb in its role and will be an asset in a conventional war to gain and maintain air supremacy. But is that capability linked to a relevant strategy? With the U.S. military engaged in simultaneous counterinsurgency campaigns, the answer is *no*. If in ten years the United States is engaged in a conventional war against a peer competitor, then the answer is *yes*. More important, that decision should not be left to the U.S. Air Force, which is likely to fund programs based on their own conception of strategic need rather than on the U.S. military's need to support strategy holistically.

The requirement for the U.S. military to be able to engage in both conventional and counterinsurgency operations needs to be recognized and not as an ad hoc decision after it is already engaged in a counterinsurgency effort. To that end, producing dual-role technology should be a priority to provide the greatest possible strategic flexibility to the U.S. military. There is not a reasonable dual role for the F-22 Raptor in counterinsurgency operations. Further, enforcing dual-role systems cannot be left to the individual services. They are far too likely to expand capabilities in evolutionary steps and within their own comfort zones. In this respect, only civilian oversight will provide the appropriate framework within which the military must structure itself and fund systems.

Sensing and information systems are the future of properly leveraged technology for a counterinsurgency conflict. Information provided directly to the lowest echelons through their own indigenous sensing systems that are under their direct control is the most appropriate way to further counterinsurgency strategy. The scope and capabilities of the sensing requirements far exceed what are needed for a line unit in a conventional conflict; however, line units in conventional conflicts can use the same systems to enhance awareness and, more important, stabilize an environment after the conventional conflict has ended. It should be assumed that any conventional victory obtained by the U.S. military will be followed by a counterinsurgency struggle while a transition to self-rule of the contested territory takes place. In such an environment, the sensing capabilities that make the operationally offensive, tactically defensive concept viable will pay enormous dividends.

The exception to assuming that a counterinsurgency effort will have to follow a conventional conflict is if the United States starts to engage in raiding strategies. Without question raiding operations would benefit greatly from sensing technologies and a heightened ability of the U.S. military to operate in small, self-sufficient units. Major legalistic issues need to be overcome if the U.S. military is going to enter into raiding and punitive operations. If the political leadership deems the operations necessary, then technology will facilitate the execution of such operations. So while technology should be tailored to implement counterinsurgency strategy and raiding strategies effectively to prepare the U.S. military for future challenges, a similar effort to make them politically acceptable should be exerted concurrently. Without both efforts, the technical advantages may be wasted.

VI. Counterinsurgency Context and Complexity

Counterinsurgency operations are extremely complex and varied. No aspect should be considered in isolation from other influences. Every facet of counterinsurgency strategy is directly linked to other facets, and every conflict is unique. While there are underlying themes and constants in counterinsurgency operations, no conflict or successful strategy in one counterinsurgency conflict should be assumed to be a guarantee of success in another counterinsurgency environment. Languages, social customs, religions, and cultures all play important roles in tailoring a counterinsurgency strategy. Moreover, in every attempt to support a host nation government through counterinsurgency practice, the security of the local populations is paramount. People cannot be counted on to support a government that cannot provide them with security. So while the constant in counterinsurgency is population security, the questions of how that security is achieved and how the U.S. military interacts with the culture need careful consideration at all levels.

A failure to tailor U.S. military behavior to local norms greatly increases the risk of alienating the populace. While it is not required to have a populace actively supporting the U.S. military—the people's inactivity is enough—no counterinsurgency force can afford to drive the people into the arms of insurgents through foolish mistakes and unnecessary insults. With every mistake potentially costing soldiers' lives, even the lowest-ranking privates need to take seriously the danger posed through thoughtless acts that may be taken as insults. Appropriate behavior of the counterinsurgency force must be a constant concern throughout the effort, which is likely to be a lengthy process.

Given the duration of counterinsurgency conflicts, the U.S. military should expect several iterations and repetitions of insurgent adaptation. Simply hoping that insurgents will not adapt their tactics and tools will result in disappointment. If the adaptation is expected, then countering it with new counterinsurgency tactics should be achievable. If the adaptation is unexpected, then developing a response may be more difficult, but innovation and problem solving are essential. Adaptation at the quickest feasible speed is an absolute requirement. If new technical capabilities are possible, funding should be prompt and implementation quickly achieved. The process of adaptation by insurgents, and countermeasures developed by the counterinsurgency forces, will continue until the insurgency can be driven back into Mao's first stage—that is, they must go into hiding.

Judging the progress of a counterinsurgency effort should not be done more than once per year. Minor fluctuations in operations with attendant small-scale attacks and ambushes by insurgents can be interpreted any way a distracted press desires. As body counts were a meaningless metric in Vietnam, so too are the numbers of American casualties or dollars spent today. Counterinsurgency efforts will always cost money and lives, but those metrics do not necessarily reflect anything about the progress of a given conflict. The only appropriate measure is whether political objectives are being met—that is, whether the insurgency is moving toward Mao's first stage of revolutionary warfare or becoming more entrenched in Mao's second stage.

Because of the difficulty in judging the progress of a counterinsurgency effort, the implication is that an administration should tell the media and the American people bluntly that attempts to micro-judge the conflict are counterproductive and that an official review will be made once a year. The press will exert tremendous pressure for regular updates on the progress of a counterinsurgency conflict; however, there is very little to be gained by making predictions about an inherently unpredictable insurgency. Foreign political actors who are required to implement changes to reduce the appeal of an insurgency must make decisions within their own time frame and political realities. Counterinsurgency forces need to show patience and restraint over a long period before the trajectory of the conflict will become evident.

Intelligence is the key tool of counterinsurgency forces. Sensing and other technical capabilities combined with HUMINT-driven intelligence not only ensure better security for the population but also are instrumental in providing appropriate civil and kinetic opportunities. Actions that are not directly driven by solid intelligence are more likely to be useless at best and potentially harmful to the effort. Maneuver battalions thrashing about the countryside did not work in Vietnam and will not work in current or future counterinsurgency efforts.

Kinetic operations are almost as important as population security objectives, especially if insurgents have safe havens in neighboring independent and sovereign states. Insurgent true believers in their cause who are, and will continue to be, irreconcilable with the host nation government need to be eliminated or incarcerated. People who are not necessarily true believers also should not be allowed to think they can join the insurgency without risk. Associated kinetic actions must be driven by solid intelligence while practicing proper operational security, moving to target areas quickly, and using precision-guided supporting fires. Without kinetic operations, the crucial component of fear to help impose political will is absent.

Counterinsurgency operations are not simply social engineering projects aimed at nation building; they are warfare in every sense. Imposing political will through the use or threat of force remains a constant in all forms of warfare. It is important that the civil aspects of counterinsurgency warfare do not become the sole focus of the effort. Liberal democratic countries do not want to emphasize the violent aspects of warfare; however, as stated, the nature of war is unchanging regardless of the ideology of the counterinsurgency forces involved.

Owing to the generally casualty-shy nature of foreign audiences and the U.S. government's inability to convince domestic audiences that warfare is a necessary evil, the United States should not count on widespread allied support in coalition-based counterinsurgency conflicts. The long-term nature of properly executed counterinsurgency efforts makes it unlikely that coalition partners will be able to last throughout the conflict. While their support may be instrumental in lending legitimacy at the inception of a counterinsurgency effort, their different levels of interoperability and combat effectiveness should not be assumed to contribute in a serious way. The United Kingdom is one of the few exceptions, but even its willingness to support U.S. military efforts has limitations.

As the U.S. military continues to introduce advanced technology into the force structure, working closely with foreign militaries will become increasingly difficult.[9] Even if a coalition partner genuinely wanted to participate fully in joint combat operations, the process will become more and more complicated to the point that their participation may become counterproductive. The implications for lacking the ability to interoperate with other militaries, and the likelihood that those foreign populations will not support long-term counterinsurgency efforts, lead to the conclusion that the only relevant role of coalition building may be as a facade of international support at the inception of the operation.

The American people are also likely to tire of an endless counterinsurgency engagement that does not produce reasonable progress over time. Insurgents are well aware of this vulnerability of the U.S. military. The American people's perception of a counterinsurgency conflict's progress is, in reality, the controlling factor. Convincing the American public that the effort is too expensive, either in lives or in treasure, represents the best chance for the insurgents to be rid of U.S. military interference and to obtain their political objectives. It is paramount that an administration present to the people, the final arbiters of whether any U.S. military effort can continue, a rational justification of the military effort.

This book has attempted to contextualize the role of technology in counterinsurgency operations, to make sense of technical trends, to address challenges facing the U.S. military, and to examine the exponential growth in technology's potential. The need for such a perspective is largely due to a lack of coherent strategic flexibility demonstrated by the U.S. military and its civilian overseers. The operationally offensive, tactically defensive concept provides a framework for the proper application of technical capabilities to provide continuity between the levels of warfare while directly supporting classic counterinsurgency practice. However, the concept is not meant to be, nor should it be interpreted as being, a silver bullet that offers a foolproof doctrine for counterinsurgency warfare. As always, the art of strategy must choose the most appropriate ways to use available means effectively to reach desired ends.

If the practitioners of politics are not versed in or remain ignorant of strategic considerations, and military professionals are unwilling or unable to appraise a conflict honestly, no amount of technology or research and development will overcome strategic shortcomings. The art of strategy will remain stunted in a technophilic environment. Congressional spending on defense will continue, jobs will be created, and the United States will keep pursuing its national interests. The United States will not likely be able to respond effectively to the non-state actor threats that will challenge it in the future. Ideally, civilian and military leaders will be more versed in strategy, given the conflicts since 2001, but strategic ineptitude has historically been the norm, not the exception, despite the many conflicts in which the United States has engaged.

The U.S. military must consider technology in both strategic and tactical counterinsurgency terms if it is going to be able to engage effectively in future counterinsurgency operations; and it should reevaluate the legalistic constraint, both domestic and international, that will prevent its use of technology to pursue and engage its enemies when counterinsurgency strategy is not viable. If the United States does so, it will be able to leverage the accumulated technical advantage that it has spent decades and vast sums of money to acquire. If it fails to do so, it will likely find itself unable to confront challenges to its security in a world where weapons of mass destruction will become more prevalent and in the hands of groups that cannot be influenced by the traditional international system of states.

Notes

Strategy and Context

Epigraph: Goldman, *Gorbachev's Challenge*, 7.

1. Jordan, *Understanding Modern Warfare*, 236.
2. Blank et al., *Low-Intensity Conflict*, 165.
3. Bull, *Anarchical Society*, 10.
4. Christian, *Nicaragua*, 234; and Carew, *Jihad!*, 278.
5. Bull, *Anarchical Society*, 8.
6. Macgregor, *Breaking the Phalanx*, 31.
7. Vickers and Martinage, "Revolution in War," 2.
8. Vickers and Martinage, "Revolution in War," 1.
9. Padover, *Living U.S. Constitution*, 11.
10. Fall, *Street without Joy*, 11.
11. Taber, *War of the Flea*, 54.
12. Poole, *One More Bridge*, 48–49.
13. CNN, "Roadside Bombs."
14. Nagl, *Learning to Eat Soup*, 10–11.
15. Thompson, *Defeating Communist Insurgency*, 124–25.
16. Schwarzkopf, *It Doesn't Take a Hero*, 119.
17. Fotion, *War and Ethics*, 8.
18. Prokosch, *Technology of Killing*, xii.
19. Burchill, *Theories of International Relations*, 71.
20. Nagl, *Learning to Eat Soup*, 18.
21. Tzu, *The Art of War*, 8.
22. Clausewitz, *On War*, 101.
23. Clausewitz, *On War*, 119.
24. Mao and Griffith, *Yu Chi Chan*, 94–114.
25. Leckie, *Wars of America*, 571.
26. Leckie, *Wars of America*, 570.
27. Clausewitz cited in Jordan, *Understanding Modern Warfare*, 44.
28. Lynn, "Patterns of Insurgency."
29. Clausewitz, *On War*, 101.
30. Petraeus and Amos, *FM 3-24: Counterinsurgency*, 1–20.
31. Taber, *War of the Flea*, 42.

32. Luttwak, *Coup d'État*, 146.

33. Taber, *War of the Flea*, 47–51.

34. Kilcullen, *Accidental Guerrilla*, 294–95.

35. Petraeus and Amos, *FM 3-24: Counterinsurgency*.

36. Galula, *Counterinsurgency Warfare*, 82.

37. Callwell, *Small Wars*, 129.

38. Callwell, *Small Wars*, 75.

39. Kitson, *Bunch of Five*, 156.

40. Scheuer, *Through Our Enemies' Eyes*, xv.

41. Kitson, *Bunch of Five*, 50–60.

42. Thompson, *Defeating Communist Insurgency*, 86.

43. Taber, *War of the Flea*, 46.

44. Taber, *War of the Flea*, 192; and Thompson, *Defeating Communist Insurgency*, 86–87.

45. Nagl, *Learning to Eat Soup*, 50.

46. Kilcullen, *Accidental Guerrilla*, 82.

47. Tucker, "Confronting the Unconventional," 1.

48. Denselow, "Agent Orange Blights Vietnam."

49. Hurley, *Taking the "Revolution,"* 6.

50. Gray, *Strategy for Chaos*, 123.

51. Friedman and Friedman, *Future of War*, 124–25.

52. Interview, Representative Larry Seaquist (D), Twenty-Sixth District, Washington State, August 19, 2008. Note that all interviews for this book were conducted by the author.

53. Friedman and Friedman, *Future of War*, 127.

54. Friedman and Friedman, *Future of War*, 124.

55. Gray, *Transformation and Strategic Surprise*, 4.

56. Yousaf and Adkin, *Afghanistan: The Bear Trap*, 184.

57. Scales, *Yellow Smoke*, 108.

58. Cecchine et al., *Army Medical Strategy*, 32.

1. The Operationally Offensive, Tactically Defensive Concept

First epigraph: Khong, *Analogies at War*, 148. *Second epigraph*: Capt. John Fry, commander, Alpha Battery, 2/12 Field Artillery, Fourth Brigade, Second Infantry Division, U.S. Army, March 14, 2008, outside of Abu Khamis, Diyala Province, Iraq.

1. Scales, *Yellow Smoke*, 109.

2. Montross, *War through the Ages*, 649.

3. Leckie, *Wars of America*, 464.

4. Scales, *Yellow Smoke*, 110–11.

5. Hastings, *Oxford Book of Military Anecdotes*, 288.

6. Billings-Yun, *Decision against War*, 24.

7. Greenfield, *Command Decisions*, 442.

8. British 1st Airborne Division Living History Association, *Brief Battle History*, 2005.

9. Mao and Griffith, *Yu Chi Chan*, 26.

10. Mao and Griffith, *Yu Chi Chan*, 31.

11. Leckie, *Wars of America*, 467.

12. Dougherty and Eidt, "Wound Ballistics," 1.

13. Patton, *War as I Knew It*.

14. Scales, *Yellow Smoke*, 109.

15. Freeman, *Lee's Lieutenants*, 42.

16. Ashbrook Center, *American Civil War Industry*.

17. Montross, *War through the Ages*, 701.

18. Friedman and Friedman, *Future of War*, 121.

19. Dupuy and Dupuy, *Encyclopedia of Military History*, 948.

20. Fall, *Hell in a Very Small Place*, 90.

21. Melton and Pawl, *Civil War Artillery Projectiles*, 26.

22. Dupuy and Dupuy, *Encyclopedia of Military History*, 823.

23. Fuller, *Conduct of War*, 168.

24. Gudmundsson, *Stormtroop Tactics*, 94.

25. Zabecki, *Steel Wind*, 16.

26. West, "Minie Balls."

27. Based on personal experience using a 1903 Springfield rifle, .30-06 caliber.

28. Friedman and Friedman, *Future of War*, 121.

29. Friedman and Friedman, *Future of War*, 122.

30. Friedman and Friedman, *Future of War*, 123.

31. Dupuy and Dupuy, *Encyclopedia of Military History*, 949.

32. Ellis and Chamberlain, *Fighting Vehicles*, 13.

33. Montross, *War through the Ages*, 700.

34. Travers, *Killing Ground*, 62.

35. Fuller, *Conduct of War*, 160.

36. Gudmundsson, *Stormtroop Tactics*, 71.

37. Fussell, *Great War*, 43.

38. Montross, *War through the Ages*, 718.

39. Paschall, *Defeat of Imperial Germany*, 78.

40. Fuller, *Conduct of War*, 166.

41. Montross, *War through the Ages*, 716.

42. Fuller, *Conduct of War*, 166.

43. Fall, *Viet-Nam Witness*, 35.

44. Fussell, *Great War*, 39.

45. Fall, *Hell in a Very Small Place*, 1.

46. Windrow, *Last Valley*, 223.

47. Fall, *Viet-Nam Witness*, 30.

48. Van, *Our Great Spring Victory*, 16–18.

49. Khong, *Analogies at War*, 149; and Khong, *Analogies at War*, 155.

50. Fall, *Viet-Nam Witness*, 138–39, 346.

51. Fall, *Street without Joy*, 40.

52. Moore and Galloway, *We Were Soldiers Once*, 199.

53. Vick et al., *Air Power*, 70–71.

54. Khong, *Analogies at War*, 152.

55. Billings-Yun, *Decision against War*, 6.

56. Department of Defense, DOD *Dictionary*.

57. Fall, *Hell in a Very Small Place*, 31.

58. Simpson, *Dien Bien Phu*, 67.

59. Dupuy and Dupuy, *Encyclopedia of Military History*, 1296.

60. Billings-Yun, *Decision against War*, 154.

61. Dupuy and Dupuy, *Encyclopedia of Military History*, 1296.

62. Fall, *Hell in a Very Small Place*, 102.

63. Windrow, *Last Valley*, 315.

64. Zumbro, *Iron Cavalry*, 460.

65. Billings-Yun, *Decision against War*, 41.

66. Fall, *Hell in a Very Small Place*, 455.

67. Billings-Yun, *Decision against War*, 135.

68. Fall, *Hell in a Very Small Place*, 342.

69. Windrow, *Last Valley*, 629.

70. Fall, *Hell in a Very Small Place*, 455.

71. Simpson, *Dien Bien Phu*, xxi.

72. Fall, *Hell in a Very Small Place*, 441.

73. Keller, "Piezoresistive Technology."

74. Windrow, *Last Valley*, 112.

75. Fall, *Hell in a Very Small Place*, 452.

76. Fall, *Hell in a Very Small Place*, 453.

77. Barker and Air Ministry, *Berlin Air Lift*, 5.

78. Windrow, *Last Valley*, 508.

79. Stratemeyer and Y'Blood, *Three Wars*, 209.

80. Fall, *Hell in a Very Small Place*, 337.

81. Fall, *Hell in a Very Small Place*, 452.

82. O'Connell, *Effectiveness of Airpower*, 70.

83. Stratemeyer and Y'Blood, *Three Wars*, 26.

84. Fuller, *Conduct of War*, 106.

85. Jordan, *Understanding Modern Warfare*, 194; and Arreguín-Toft, *How the Weak Win*, 40.

86. Friedman and Friedman, *Future of War*, 234.

87. Macgregor, *Breaking the Phalanx*, 52.

88. Scales, *Yellow Smoke*, 110.

89. Winters, *Battling the Elements*, 61.

90. Kinross, *Clausewitz and America*, 63.

91. Kissinger and Luce, *White House Years*, 990.

92. Thompson and Frizzell, *Lessons of Vietnam*, 138.

93. Winters, *Battling the Elements*, 63, 69.

94. Winters, *Battling the Elements*, 70.

95. Starry, *Armored Combat in Vietnam*, 116.

96. Winters, *Battling the Elements*, 71; and Thompson and Frizzell, *Lessons of Vietnam*, 179, 141.

97. Kissinger and Luce, *White House Years*, 1374.

98. Winters, *Battling the Elements*, 71.

99. Winters, *Battling the Elements*, 71.

100. Khong, *Analogies at War*, 173.

101. O'Connell, *Effectiveness of Airpower*, 54.

102. Thompson and Frizzell, *Lessons of Vietnam*, 85.

103. Thompson and Frizzell, *Lessons of Vietnam*, 76.

104. Shay, *Achilles in Vietnam*, 19.

105. Friedman and Friedman, *Future of War*, 237.

106. U.S. Army, *Introduction*, 1:3–2; and Kinross, *Clausewitz and America*, 160.

107. For an examination of technological innovations, see chapter 4.

108. Interview, Anthony Nelson, Tampa FL, March 17, 2007.

109. Marshall, *Soldier's Load*, 5.

110. Leckie, *Wars of America*, 468.

111. Fall, *Hell in a Very Small Place*, 455.

112. Thompson and. Frizzell, *Lessons of Vietnam*, 137.

113. Thompson, *Defeating Communist Insurgency*, 121.

114. Interview, Nelson.

115. Thompson, *Defeating Communist Insurgency*, 112.

116. Schmidtchen, *Eyes Wide Open*, 4.

117. Based on personal experience, combat patrols, Diyala Province, Iraq, March–April 2008.

118. Ricks, *The Gamble*, 203.

119. Personal experience, combat patrols, Diyala Province, Iraq, March–April 2008.

120. Taber, *War of the Flea*, 192.

121. Lesser et al., *Countering the New Terrorism*, 35.

122. Ricks, *The Gamble*, 304.

123. Galula, *Counterinsurgency Warfare*, 7–8.

124. Nagl, *Learning to Eat Soup*, 222.

125. Leckie, *Wars of America*, 467.

126. Mao and Griffith, *Yu Chi Chan*, 26.

127. Fall, *Hell in a Very Small Place*, 51.

128. Macgregor, *Breaking the Phalanx*, 73.

129. Friedman and Friedman, *Future of War*, 301.

130. Interview, Nelson.

131. Handel, *Sun Tzu and Clausewitz*, 7.

2. Counterinsurgency in Iraq

Epigraph: Aurelius, *The Great Books*, 267.

1. Creveld, *Transformation of War*, 20, 29.

2. Batiste, CBS interview, *Face the Nation*.

3. Interview, Anthony Nelson, Tampa FL, March 17, 2007.

4. Central Intelligence Agency, "Middle East: Iraq."

5. DeLong, *Inside CENTCOM*, 121.

6. Gaddis, *Surprise, Security*, 21.

7. Scheuer, *Through Our Enemies' Eyes*, 251, 252.

8. Chandrasekaran, *Imperial Life*, 37.

9. Haddick, "This Week at War."

10. Johns, "Crimes of Saddam Hussein."

11. DeLong, *Inside CENTCOM*, 83.

12. Hashim, *Iraq's Sunni Insurgency*, 28.

13. Hashim, *Iraq's Sunni Insurgency*, 29.

14. Woodward, *Plan of Attack*, 82.

15. Interview, Representative Larry Seaquist (D), Twenty-Sixth District, Washington State, August 19, 2008.

16. Woodward, *U.S. Foreign Policy*, 62, 63.

17. Interview, Central Command science adviser Earl Rubright, Tampa FL, January 1, 2002.

18. U.S. Army, "Operational Army."

19. Gordon and Trainor, *The Generals' War*, ix; and Belasco, "Troop Levels."

20. Hoar, "Why Aren't There Enough?"; and Schmidt, "Pentagon Contradicts."

21. DeLong, *Inside CENTCOM*, 123–24.

22. Wright, *Generation Kill*, 251.

23. Woodward, *Plan of Attack*, 81.

24. Cohen and Keaney, "Gulf War Air Power," 225.

25. Interview, Col. Michael "Coyote" Smith, USAF, Reading, UK, April 16, 2010.

26. Rosenau, "Special Operations Forces."

27. Federation of American Scientists (FAS), "Laser Guided Bombs."

28. Interview, Smith.

29. Interview, Smith.

30. Interview, Col. Steve Latchford, USAF, Tampa FL, June 3, 2008.

31. Interview, Smith.

32. Interview, Capt. John Fry, USA, A 2-12 FA 4/2 ID, FOB Warhorse, Iraq, March 14, 2008.

33. Anderson, "Air and Missile Defense," 40–47.

34. Interview, Maj. Mike Garcia, USA, HHQ 4/2ID, FOB Warhorse, Iraq, February 28, 2008.

35. Lesser et al., *Countering the New Terrorism*, xii.

36. Interview, Garcia.

37. Interview, Master Sergeant Wallace, USA, A 2-12 FA 4/2ID, COB DMC, Iraq, March 16, 2008.

38. Interview, Sergeant Farmer, USA, A 2-12 FA 4/2ID, COB DMC, Iraq, March 17, 2008.

39. Interview, Farmer.

40. Personal observation at Abu Khamis, Diyala Province, Iraq, March 12, 2008.

41. Interview, Translator Mahmmod Dadow, attached A 2-12 FA 4/2ID, Abu Khamis, Diyala Province, Iraq, March 12, 2008.

42. Ricks, *The Gamble*, 5–6.

43. Ricks, *The Gamble*, 5–6.

44. Dupont, "Protecting Those Who Protect Us."

45. Interview, Lt. Col. Andre Fallot, USA, brigade surgeon, 4/2ID, FOB Warhorse, Iraq, April 1, 2008.

46. Naylor, *Not a Good Day*, 355.

47. Leading Technologies Composites, printing on front trauma plates for use in level 3A vests, 2008.

48. Interview, Fallot; and interview, Corporal Smith, USA, Troop C 2/1 CAV 4/2ID, Khan Bani Saad, Iraq, April 4, 2008.

49. Interview, Corporal Smith.

50. Interview, Staff Sgt. Ryan Gery, USA, A CO 2-23 INF 4/2ID, FOB Warhorse, Iraq, March 23, 2008.

51. Interview, Maj. Tim Cass, USA, brigade physical therapist, HHC 4/21D, FOB Warhorse, Iraq, April 3, 2008.

52. Tally taken from the wall in the brigade's tactical operations center, FOB Warhorse, Iraq, as of March 28, 2008. The author counted them in person.

53. Personal experience in U.S. Army, B CO 3BAT 20 SFG (A), 1992.

54. Interview, Staff Sgt. Michael Thomas, USA, HHC 4/21D, FOB Warhorse, Iraq, March 19, 2008.

55. Carollo, "Deadly Price Paid"; and BAE Systems, "M1114 Up-Armored HMMWV."

56. Interview, Maj. Patrick Mackin, USA, S2 4/21D, FOB Warhorse, Iraq, February 15, 2008.

57. Crosby, "MRAPs Hits the Streets," 4.

58. Crosby, "MRAPs Hits the Streets," 5.

59. Interview, Mackin.

60. Interview, Mackin.

61. Interview, Maj. Mathew Moore, USA, executive officer, 2/23 IN 4/21D, FOB Warhorse, Iraq, March 4, 2008.

62. Owens, *Lifting the Fog*, 14.

63. Interview, Capt. David Siler, USA, D Troop 2/1 CAV 4/2 ID, FOB Warhorse, Iraq, March 23, 2008.

64. Interview, Siler.

65. Interview, Siler.

66. Interview, Garcia.

67. Interview, Lt. Col. Ricardo Love, USA, commander, 1/38 IN 4/21D, FOB Warhorse, Iraq, March 29, 2008.

68. Interview, Siler.

69. Interview, Siler.

70. Interview, Garcia.

71. Interview, Petty Officer First Class (PO1) Sean Mulligan, USN (Attached), HHC 4/21D, FOB Warhorse, Iraq, March 1, 2008.

72. Interview, Garcia, April 3, 2008.

73. Personal experience, air reaction force mission with A Troop 2/1 CAV 4/2 ID, Diyala Province, Iraq, April 2, 2008.

74. Interview, Garcia, April 3, 2008.

75. Interview, Maj. Thomas Rider, USA, provost marshal, HHC 4/21D, FOB Warhorse, Iraq, April 4, 2008.

76. Blank, *Responding*, 36.

77. Waxman, "International Law," 8.

78. Personal experience, air reaction force mission with A Troop 2/1 CAV 4/2 ID, Diyala Province, Iraq, April 2, 2008.

79. Personal experience, air reaction force mission with A Troop 2/1 CAV 4/2 ID, Diyala Province, Iraq, April 2, 2008.
80. Operation Iraqi Freedom, "Diyala."
81. Interview, Col. John Lehr, USA, commander, 4/2ID, FOB Warhorse, Iraq, April 4, 2008.
82. Interview, Mackin.
83. Interview, Mackin.
84. Department of Defense, "Start of 'Arrowhead Ripper.'"
85. Personal experience, FOB Warhorse, Iraq, Spring 2008.
86. Interview, Lehr.
87. Fox News, "Transcript: Gen. David Petraeus."
88. Interview, Lehr.
89. Interview, Rider.
90. Interview, Mackin.
91. Interview, Capt. Phillip Mundweil, USA, HHC 1/38 IN 4/2 ID, FOB Warhorse, Iraq, February 26, 2008.
92. Interview, Mundweil.
93. Interview, Lt. Col. Marshall Dougherty, USA, commander, 2/1 CAV 4/2 ID, Khan Bani Saad, Iraq, April 3, 2008.
94. Interview, Lt. Joseph Covey, USA, platoon leader, A CO 4/9 IN 4/2 ID, Old Baquba, Iraq, March 17, 2008.
95. Interview, Mundweil.
96. Benjamin and Simon, *Next Attack*, 54–57.
97. Interview, Mundweil, February 27, 2008.
98. Interview, Mundweil, February 27, 2008.
99. Fox News, "Transcript: Gen. David Petraeus."
100. Public Broadcasting Service, "Interview: Lt. Col. Ernest 'Rock' Marcone."
101. Cockburn, "Sunni vs. Shia."
102. Nagl, *Learning to Eat Soup*, 4–6.
103. Interview, Seaquist.
104. Schmitt, "Defense Reorganization"; and interview, Seaquist.
105. Ricks, *The Gamble*, 25.
106. Andrews, "Envoy's Letter."
107. Interview, Covey.
108. Kikuchi, "Iraq Troop Surge."
109. Kikuchi, "Iraq Troop Surge."
110. Interview, Patrick Baz, Agence France-Presse photographer, FOB Caldwell, Iraq, March 20, 2008.
111. Barnard, "Marines See."

112. Barnard, "Marines See."

113. Interview, Baz.

114. Interview, Baz.

115. Michaels, "Petraeus Strategy."

116. Interview, Mackin.

117. Interview, Mackin.

118. Personal experience, COB DMC, Iraq, March 2008.

119. Personal experience, around COB DMC, Iraq, March–April 2008.

120. United States Army Special Operations Command, "Brief History."

121. Carden, "Iraqi Forces Improve."

122. Carden, "Iraqi Forces Improve."

123. Kilcullen, *Accidental Guerrilla*, 126.

124. Batty, "British Forces."

125. Keiler, "Who Won."

126. Personal experience, Abu Khamis, Iraq, March 14, 2008.

127. Interview, Col. Steve Latchford, USAF, Tampa FL, July 27, 2008.

128. Personal experience, Diyala Province, Iraq, February–April 2008.

129. Personal experience, Diyala Province, Iraq, February–April 2008.

130. Kramer, "4th Brigade's Deployment."

3. Limits of Politically Correct Doctrine

Epigraph: Thucydides, *The Peloponnesian War*, 506–8.

1. Galula, *Counterinsurgency Warfare*, 81–83.

2. Petraeus and Amos, *FM 3-24: Counterinsurgency*, 1–20.

3. Petraeus and Amos, *FM 3-24: Counterinsurgency*, 1–3.

4. *The Economist, Pocket World in Figures*, 154.

5. Department of Defense, "Casualty Reports."

6. *The Economist, Pocket World in Figures*, 36.

7. Brady, "Evangelical Chaplains."

8. Viola, "What Is the Proper Role."

9. Sarin and Dvoretsky, *Afghan Syndrome*, 124–25.

10. Lang, "What Iraq Tells U.S."

11. Lambeth, *Air Power*, 33.

12. Lambeth, *Air Power*, 104.

13. Bowman and Siegel, "Examining U.S. Goals."

14. Urban, *War in Afghanistan*, 1.

15. McFadden and Kannampilly, "Clinton Nears Decision."

16. Finel, "An Alternative to COIN."

17. Arnold, *Afghanistan*, 98–99.

18. Clausewitz, *On War*, 109, 119.

19. Gray, *The Sheriff*, 27.

20. Thistlewaite and Katulis, "How to Make Afghanistan War."

21. Knight, "Amnesty International."

22. International Committee of the Red Cross, "Convention (IV)," Article 33.

23. Nossal, "No Exit," in Ehrhart and Pentland, *Afghanistan Challenge*, 163.

24. Callwell, *Small Wars*, 75–77.

25. Agence France-Presse, "Israel, Hamas."

26. Yoo, *War by Other Means*, 8–9.

27. Kull, "Negative Attitudes."

28. Monks, *Soviet Intervention*, 15–17.

29. Galeotti, *Soviet Union's Last War*, 2.

30. Rubin, *Search for Peace*, 80.

31. Yousaf and Adkin, *Afghanistan: The Bear Trap*, 177.

32. Newell and Newell, *Struggle for Afghanistan*, 134.

33. Girardet, *Afghanistan*, 33.

34. Rais, *War without Winners*, 112.

35. Yousaf and Adkin, *Afghanistan: The Bear Trap*, 59–61.

36. Girardet, *Afghanistan*, 81–83.

37. Girardet, *Afghanistan*, 80.

38. Newell and Newell, *Struggle for Afghanistan*, 149.

39. Yousaf and Adkin, *Afghanistan: The Bear Trap*, 34.

40. Personal conversation with Professor Colin Gray, Reading, UK, Fall 2008.

41. Yousaf and Adkin, *Afghanistan: The Bear Trap*, 182.

42. Rais, *War without Winners*, 102; and Urban, *War in Afghanistan*, 157.

43. Yousaf and Adkin, *Afghanistan: The Bear Trap*, 180.

44. Carew, *Jihad!*, 270.

45. Personal experience with the SA-7 missile system.

46. Sarin and Dvoretsky, *Afghan Syndrome*, 66.

47. Personal experience with the Stinger missile system.

48. Keaney, "Lessons Unlearned."

49. Yousaf and Adkin, *Afghanistan: The Bear Trap*, 56.

50. Rashid, *Descent into Chaos*, 267.

51. Welch, *Crimes of Power*, 113.

52. Lock Pullan, "US Military Strategy," 166–67.

53. Cordesman, *Lessons of Afghanistan*, 33, 37.

54. Cordesman, *Lessons of Afghanistan*, 66–99.

55. Cohen and Agiesta, "Poll of Afghans."

56. McMichael, "Hold Fire."

57. Cordesman, *Lessons of Afghanistan*, 6, 7.

58. Rhem, "Bush."

59. Cordesman, *Lessons of Afghanistan*, 6.

60. Arnold, *Afghanistan*, 97.

61. Bradley, *Vietnam at War*, 172–73.

62. Tanner, *Afghanistan*, 275–76.

63. Davidson, "Sophisticated about Corruption?"

64. Ministry of Foreign Affairs, "Ethnic Groups."

65. Tanner, *Afghanistan*, 298–99.

66. Goodson, *Afghanistan's Endless War*, 70.

67. Prados, *Vietnam*, 530–31.

68. Legal Information Institute, U.S. Code, Title 10–Armed Forces.

69. BBC, "World Citizens."

70. Personal interview, Joshua Harrow, attorney at law, Tampa, Florida, August 3, 2010.

71. Tolz, "Independence for Russian Minorities," 215.

4. Technology and Counterinsurgency Strategy

Epigraph: Ruelle, *Chance and Chaos*, 3.

1. Franks, *American Soldier*, 252.

2. Dupuy, *Evolution of Weapons*, 326–27.

3. Vickers and Martinage, "Revolution in War," 2.

4. United States General Accounting Office, "Military Transformation."

5. U.S. Army, "Operational Army"; and Nardulli and McNaugher, "The Army," 101.

6. Matsumura et al., "Army after Next," 3.

7. Wilmoth, "False-Failed Innovation," 57.

8. Feickert, "Army's Future Combat System," 6.

9. Steele, "Military Reengineering," 59–60. Italics in the original.

10. NASA, "Innovative Partnership Program."

11. Comstock and Lockney, "NASA's Legacy."

12. Perez-Davis et al., "Energy Storage."

13. General Dynamics Canada, "Fire Control"; Carl Zeiss Optronics, "Eyesafe Laser Rangefinder"; and Chait, Lyons, and Long, "Critical Technology Events," 14.

14. Amos, "Lasers Scan Future Possibilities."

15. Library of Congress, "Modern Automobile Manufacturing."

16. Shinseki, "Concepts," ii.

17. Browning Firearms, "Browning BAR History."

18. American Firearms Institute, "Important Dates."

19. Boeing, "F-22 Raptor Technical Specs."

20. Boeing, "History: B-29 Superfortress."

21. Buzan, *Introduction to Strategic Studies*, 98.

22. Lock-Pullan, "U.S. Military Strategy," 166.

23. Lockheed Martin, "LOSAT Successfully."

24. Lockheed Martin, "Lockheed Martin Receives"; and Lockheed Martin, "LOSAT Defeats."

25. Lockheed Martin, "LOSAT Defeats."

26. Kristol, "Defense Secretary We Have."

27. Satterfield, "Long and Winding."

28. Lock-Pullan, "U.S. Military Strategy," 168–69.

29. U.S. Army, "5th Special Forces Group."

30. Haas, *War of Necessity*, 276.

31. Rothstein, *Afghanistan and the Troubled Future*, 10.

32. U.S. Army, "Special Forces: Training Overview."

33. Paddock, *U.S. Army Special Warfare*, 146.

34. Tucker and Lamb, *Special Operations Forces*, 151.

35. U.S. Army, "Special Forces: The Missions."

36. Alberts, Garstka, and Stein, *DoD C4ISR Cooperative Research*, 2–3.

37. U.S. Marine Corps, *Command and Control*, 51.

38. Bially, "Lifting the Fog of War," 87–90.

39. West, *Strongest Tribe*, 102–3; and Ricks, *The Gamble*, 12–13.

40. U.S. Army, "Chapter 5: Soldier's Load."

41. Drummond, *Velocity Management*, 25.

42. Associated Press, "Troops Stationed."

43. Paget, *Counterinsurgency Operations*, 35, 36.

44. Personal experience, near FOB Caldwell, Diyala Province, Iraq, March 2008.

45. Personal experience, FOB Caldwell, Diyala Province, Iraq, April 2008.

46. Personal experience, FOB Caldwell, Diyala Province, Iraq, April 2008.

47. Frankland, *Encyclopedia of Twentieth-Century Warfare*, 375.

48. Brook, "Hulking, Tough MRAP."

49. Personal experience, Abu Khamis, Diyala Province, Iraq, March 2008.

50. Fox, "JSTARS."

51. Boeing, "Defense, Space and Security."

52. Fredricksson, "Space Power."

53. Brogan, "Marine Corps."

54. ABC News, "Improved Body Armor."

55. Callier, "Artificial Spider Silk."

56. Ucko, *New Counterinsurgency Era*, 11.

57. Mahalik, *Sensor Networks*, 1.

58. Morrison, "Sonic Silver Bullet."

59. Cacciottolo, "Biometric Science."

60. Webster, "Biometric ATM."

61. Slack, "Ministers Accused"; and Coimbatore et al., "Sensing Biological," in Kendall et al., *Advances*, 164.

62. Gillespie, *Algeria*, 146.

63. Chandrasekaran, "In Kandahar."

64. Prados, *Hidden History*, 5, 150–51.

65. Levinger, "Taking Tora Bora."

66. Thro, *Robotics*, 30.

67. Brower, "Terrorist Threat," in Byrnes, *Advances in Sensing*, 49.

68. Staugaard, *Robotics and AI*, 10–11.

69. Hansen et al., "Dialect Analysis and Modeling."

70. Harrington, "Patent Application Title."

71. Personal interview, Col. Stephen Latchford, USAF, Tampa, Florida, August 17, 2008.

72. Paget, *Counterinsurgency Operations*, 171.

73. Dalloz, *War in Indo-China*, 172.

74. Wailes and Harrington, "Guided Parafoil," 60.

75. Global Security, "Military: MC-130P Combat Shadow."

76. Interview, Lt. Col. Andre Fallot, USA, brigade surgeon, HHQ 4/2ID, FOB Warhorse, Iraq, March 23, 2008.

77. Johnson and Cecchine, *Medical Risk*, 23.

78. Interview, Fallot.

79. Dougherty, "Sabari."

80. Shinwari, "Pakistanis Reopen."

81. Taylor, *Air Force Pilot Shortage*, 2.

82. Interview, Capt. David Siler, USA, D Troop 2/1 CAV 4/2 ID, FOB Warhorse, Iraq, March 23, 2008.

83. Drezner, *Innovative Management*, 35.

84. Krishnan, *Killer Robots*, 4.

85. Lehner, *Artificial Intelligence*, 162.

86. Nikutta, "Artificial Intelligence," in Din, *Arms and Artificial Intelligence*, 130.

87. Bergen, Doherty, and Bellen, "Public Opinion."

88. Personal experience in U.S. Army Special Forces, Ft. Bragg NC.

89. Alexander, *Cost Benefits*, 37.

90. Henry, "UAS VideoTerminal."

91. *Jane's Electro-Optic Systems*, "BAE Systems AN/PAS-13C-V."

92. Shank, *Online Learning Idea Book*, 10–12.

93. Horacio Franco, SRI Systems, December 19, 2005.

94. Ballard, *Fighting for Fallujah*, 5.

95. Clark, *Winning Modern Wars*, 166–67.

96. Personal experience in U.S. Army Special Forces.

97. Personal experience in U.S. Army Special Forces.

98. Silicon Solar, "Solar Battery Chargers," Product Line, 1999, http://www.silicon
solar.com/solar-battery-chargers.html.

99. West, *Strongest Tribe*, 61.

100. West, *Strongest Tribe*, 10.

101. Bailey, *Field Artillery and Firepower*, 510.

102. Mrozek, *Air Power*, 92.

103. Harrington and Doucette, "Army after Next."

104. Blankenship, "Battlefield Ordnance Awareness."

105. Global Security, "Target Analysis."

106. Blankenship, "Battlefield Ordnance."

107. UPS, "UPS Worldport Facts," April 2010, http://www.pressroom.ups.com/Fact
+Sheets/UPS+Worldport+Facts.

108. UPS Pressroom, "Tech Driven 'Worldport' Ensures Reliability and Capacity for
Customers Worldwide," September 27, 2002, http://web.archive.org/web/2002
1106051029/http://pressroom.ups.com/pressreleases/archives/archive/0,1363
,4170,00.html.

109. UPS, "Shipping Support," November 2010, http://www.ups.com/content/us/en
/resources/ship/index.html?WT.svl=SubNav.

110. Anderson, "FedEx Wants UAVs."

111. Jackson and Haddox, "Phantom Eye."

112. Lombardo, "Army to Field."

113. Personal experience, FOB Warhorse, Diyala Province, Iraq, March 2008.

114. Dombrowski and Gholz, *Buying Military Transformation*, 61.

115. Mrozek, *Air Power*, 117.

116. Hirsch and Oakley, *Somalia*, 127–28.

117. Cecchine, *Army Medical Support*, 18.

118. Northrup Grumman Aerospace Systems, "MQ-8B Fire Scout."

119. Dubner, "Would You Fly?"

120. National Research Council, *Decadal Survey*, 48.

121. Loeb, "Rumsfeld 'Untracks' Crusader."

122. Owens and Dumbrell, *America's "War on Terrorism,"* 5.

123. Gelpi, Feaver, and Reifler, *Paying the Human Costs*, 237.

124. Margolis, *War at the Top*, 35.

125. Margolis, *War at the Top*, 34.
126. Peterson, *History of Body Armor*, 61.
127. Swartz and Iwata, "Invented to Save Gas."
128. Rincon, "Super-Strong Body Armor."
129. Ricks, "Rumsfeld Gets Earful."

5. Issues Related to Counterinsurgency Warfare

Epigraph: Law and Mol, *Complexities*, 1.
1. Blaufar, *Counterinsurgency Era*, 4.
2. Junger, *War*, 254.
3. Cubbage, "Intelligence and the Tet Offensive," in Errington and McKercher, *Vietnam War as History*, 112.
4. Buttinger, *Vietnam: A Political History*, 504.
5. Gilbert, *Vietnam War*, 15.
6. Hung, "Vietnam War," in Grinter and Dunn, *American War*, 19.
7. Steinhauer, "CNN Poll."
8. Thompson and Frizzell, *Lessons of Vietnam*, 118.
9. Putsay, *Counterinsurgency Warfare*, 31.
10. Putsay, *Counterinsurgency Warfare*, 31–32.
11. Jones, "Obama's Afghanistan Deadline."
12. Global Security, "OIF-Iraq Significant Activities."
13. Elliot-Bateman, *Defeat in the East*, 141.
14. Kline, *Colombia*, 57.
15. Fadel, "As U.S. Scales Back."
16. Buttinger, *Vietnam: A Dragon Embattled*, 2:752.
17. Paddock, *U.S. Army Special Warfare*, 159.
18. Personal observation in Diyala Province, Iraq, February–April 2008.
19. Darley, "Clausewitz's Theory."
20. Burchett, *Furtive War*, 26.
21. Valeriano and Bohannan, *Counter-Guerrilla Operations*, 186.
22. Fall, *Street without Joy*, 244; and Marshall, "Big Guns Not Answer," in Fishel, *Vietnam*, 464.
23. Sorley, *A Better War*, 20.
24. Sorley, *A Better War*, 21; and Fall, *Street without Joy*, 76–78.
25. Krepinevich, *The Army and Vietnam*, 198–99.
26. Petraeus and Amos, *FM 3-24: Counterinsurgency*, 1–27.
27. Krepinevich, *The Army and Vietnam*, 76.
28. Lambeth, *Air Power*, 110.
29. Lambeth, *Air Power*, 129.

30. Naylor, *Not a Good Day*, 30.

31. Gebhardt, "Eyes behind the Lines."

32. Junger, "Farewell to Korengal."

33. Paddock, *U.S. Army Special Warfare*, 127.

34. Krepinevich, *The Army and Vietnam*, 122.

35. U.S. Army, *Introduction*, 7–5.

36. McCurry, "Rice Says Sorry."

37. Florer, "New Phase"; and U.S. Army, "Frequently Asked Questions."

38. Personal experience as a U.S. Army infantryman.

39. U.S. Army, "Frequently Asked Questions."

40. U.S. Army, *Introduction*, 1–6.

41. Personal experience in U.S. Army Special Forces.

42. Straub, "Female Combat Medics."

43. Pierson, "They're Still Women," in Isaksson, *Women and the Military System*, 30.

44. Borch, "Military Law," in De Groot and Peniston-Bird, *A Soldier and a Woman*, 338.

45. Quester, "The Problem," in Goldman, *Female Soldiers*, 224.

46. Herper, "New Portable Drug Test."

47. "Napoleon Bonaparte Quotations," ThinkExist.com.

48. Interview, Steve Blocker, sensor integration, BCI Technologies, Tampa FL, August 17, 2010.

49. Gerring, "Vietnam Analogy," in Errington and McKercher, *Vietnam War as History*, 10.

50. Cordovez and Harrison, *Out of Afghanistan*, 189–99.

51. Gordon and Lehren, "Leaked Reports."

52. Kahib, *Intervention*, 280.

53. Brook, "MRAPs Can't Stop."

54. Yousaf and Adkin, *Afghanistan: The Bear Trap*, 187.

55. BAE Systems, "BAE Systems/U.S. Navy TADIRCM."

56. Northrup Grumman, "AN/AAQ-24(V) Directional Infrared Countermeasures."

57. McAlster, *Vietnam*, 150; Smith, *Cambodia's Foreign Policy*, 162–63; Rubin, *Fragmentation of Afghanistan*, 181–82; Roggio, "Iranian-Backed Shia Terror Group"; and RAND, "Taliban's Sanctuary Bases."

58. Caldwell and Tan, *Cambodia*, 149–50.

59. Martone, "Questions over Afghan Arabs."

60. Starr, *War Coalitions*, 23–24

61. Woodward, *Plan of Attack*, 162.

62. Council on Foreign Relations, "What Is the Coalition?"

63. Woodward, *Plan of Attack*, 285.

64. Ehrhart and Pentland, *Afghanistan Challenge*, 2.

65. BBC, "Afghan Ambush."

66. Naylor, *Not a Good Day*, 275.

67. Chesser, "Afghanistan Casualties," 1–2.

68. Penfold, "German Chancellor."

69. BBC, "NATO's Afghan Forces."

70. Bellamy, *Evolution*, 34.

71. De Leon, "Close NATO's 'Capability Gap.'"

72. North Atlantic Treaty Organization (NATO), "What Does NATO Do?"

73. Gray, *The Sheriff*, 79.

74. Saad, "Americans' Opinion of UN."

75. Woodward, *Plan of Attack*, 180–81.

76. MacAskill and Borger, "Iraq War Was Illegal."

77. United Nations, "Convention on the Prohibition of the Use"; Kirby, "Depleted Uranium Ban Demanded"; United Nations, "Convention on Prohibitions or Restrictions"; Simonite, "Robot Arms Race"; BBC, "Forum Seeks to Ban"; Federation of American Scientists, "Non-lethal and Riot Control Agents"; and Monbiot, "Behind the Phosphorus Clouds."

78. Associated Press, "Taliban Urge Fighters."

79. U.S. Army, "A Leader's Guide."

80. Damphousse and Smith, "The Internet," in Kushner, *Future of Terrorism*, 216–17.

81. Reeve, *New Jackals*, 214.

82. Kepel and Milelli, *Al Qaeda*, 26–27.

83. Wells, "Al-Jazeera Accuses U.S."

84. Personal interview, Col. Michael Smith, USAF, Reading, UK, September 15, 2009.

85. Lawson, "Pakistan's Islamic Schools."

86. Lawson, "Pakistan's Islamic Schools."

87. English, *Terrorism*, 119.

88. Combs, "The Media," in Forest, *Teaching Terror*, 142.

89. Molnar, "Human Factors."

90. Associated Press, "Arafat Horrified."

91. World Public Opinion, "International Poll."

92. Patai, *The Arab Mind*, 260.

93. Gibbons, "Representing the Real," 114.

94. Carlstorm, "'Crazy Horse.'"

95. Harrison, "Americans Turn to Al-Jazeera."

96. Carroll, "Slim Majority."

97. Walt, "No 'Do-Overs.'"

98. Goldberg, *Arrogance*, 9.

99. CNN, "White House."

100. Sturgill, *Low-Intensity Conflict*, 91.

101. Department of Defense, "Operation Enduring Freedom."

102. Tyson, "Sharp Drop."

103. Central Intelligence Agency, "Country Comparison."

104. Bohan, "Obama Says."

105. Woodward, *Obama's Wars*, 290.

106. Baker and Glasser, *Kremlin Rising*, 35.

Implications

Epigraphs: Hansen, *The Civil War*, 441.

1. Komarow, "Unexpected Insurgency."

2. "Pelosi Says No," CNN.

3. Walker, "DoD Transformation."

4. Bender and Baron, "Fewer High-Quality Army Recruits."

5. *Wall Street Journal*, "Al Qaeda in Africa."

6. Klein, "The Pashtun War."

7. "US Would Not 'Admit,'" BBC.

8. United States General Accounting Office, "Combat Air Power."

9. Millen, "Tweaking NATO."

Bibliography

ABC News. "Improved Body Armor and Medicine Saves Lives in Iraq," January 30, 2006. http://abcnews.go.com/gma/OnCall/story?id=1556540.

Agence France-Presse. "Israel, Hamas Take Gaza War into Cyberspace," December 1, 2009. http://www.abs-cbnnews.com/technology/01/12/09/israel-hamas-take-gaza-war-cyberspace.

Alberts, David, John Garstka, and Frederick Stein. *DoD C4ISR Cooperative Research Program Network Centric Warfare: Developing and Leveraging Information Superiority.* Washington DC: CCRP Publication Series, 2000.

Alexander, Arthur. *The Cost Benefits of Reliability in Military Equipment.* Santa Monica: RAND, 1988.

Al Jazeera. "Taliban Takes Over Afghan Valley: Fighters Claim Victory after U.S. Forces Withdraw from Korengal Valley Base," April 19, 2010. English.aljazeera.net/news/asia/2010/04/20104196826856839.html.

American Firearms Institute. "Important Dates in Gun History," 2008. http://www.americanfirearms.org/history.php.

Amnesty International. "Suffocating Gaza—the Israeli Blockade's Effects on Palestinians," June 1, 2010. http://www.amnesty.org/en/news-and-updates/suffocating-gaza-israeli-blockades-effects-palestinians-2010-06-01.

Amos, Jonathan. "Lasers Scan Future Possibilities." BBC, May 12, 2010. http://news.bbc.co.uk/2/hi/science/nature/8675972.stm.

Anderson, Charles A. "Air and Missile Defense: Operation Iraqi Freedom." *Army* 54 (January 2004): 40–47.

Anderson, Chris. "Fred Smith: FedEx Wants UAVs." *DIY Drones*, February 12, 2009. http://diydrones.com/profiles/blogs/fred-smith-fedex-wants-uavs.

Andrews, Edmund. "Envoy's Letter Counters Bush on the Dismantling of the Iraqi Army." *New York Times*, September 4, 2007. http://www.nytimes.com/2007/09/04/washington/04bremer.html.

Arnold, Anthony. *Afghanistan: The Soviet Invasion in Perspective.* Stanford CA: Hoover Press, 1985.

Arreguín-Toft, Ivan. *How the Weak Win Wars: A Theory of Asymmetric Conflict.* Cambridge Studies in International Relations 99. New York: Cambridge University Press, 2005.

Ashbrook Center for Public Affairs. *American Civil War Industry.* Ashland University, Ashland OH, 2008.

Associated Press. "Arafat Horrified by Attacks, but Thousands of Palestinians Celebrate; Rest of World Outraged." Fox News, September 12, 2001. http://www.foxnews.com/story/0,2933,34187,00.html.

———. "Taliban Urge Fighters to Avoid Killing Civilians in New Code of Conduct to Win Hearts and Minds." Fox News, August 3, 2010. http://www.foxnews.com /world/2010/08/03/taliban-urge-fighters-avoid-killing-civilians-new-code-conduct-win-hearts-minds/.

———. "Troops Stationed in Iraq Turn to Gaming." MSNBC, January 3, 2005. http:// www.msnbc.msn.com/id/6780587/.

———. "U.S. Pulls out of Afghanistan's 'Valley of Death.'" Fox News, April 14, 2010. http://www.foxnews.com/world/2010/04/14/exits-afghanistans-valley-death/.

Aurelius, Marcus. *The Great Books: Meditations*. Chicago: William Benton, 1952.

BAE Systems. "BAE Systems/U.S. Navy TADIRCM Clears Major Development Milestone." BAE Systems news release, January 31, 2002. http://www.baesystems .com/Newsroom/NewsReleases/2002/press_310120021.html.

———. "M1114 Up-Armored HMMWV," 2009. http://www.baesystems.com/BAEProd /groups/public/documents/bae_publication/bae_pdf_mps_m1114.pdf.

Bailey, J. B. A. *Field Artillery and Firepower*. Annapolis: Naval Institute Press, 2004.

Baker, Aryn. "Pakistan: Behind the Waziristan Offensive." *Time*, October 18, 2009. http://www.time.com/time/world/article/0,8599,1930909,00.html.

Baker, Peter, and Susan Glasser. *Kremlin Rising*. New York: Scribner, 2005.

Ballard, John. *Fighting for Fallujah: A New Dawn for Iraq*. London: Praeger Security International, 2006.

Barker, Dudley, and Air Ministry, Central Office of Information. *Berlin Air Lift*. London: His Majesty's Stationery Office, 1949.

Barnard, Anne. "Marines See Little Progress in Rebel-Controlled Fallujah." *Boston Globe*, September 16, 2004. http://www.boston.com/news/world/articles/2004 /09/16/marines_see_little_progress_in_rebel_controlled_fallujah/.

Barnett, Roger W. *Asymmetrical Warfare: Today's Challenge to U.S. Military Power Issues in Twenty-First Century Warfare*. Washington DC: Brassey's, Inc., 2003.

Batiste, John. Interview. *Face the Nation*. CBS, April 23, 2006. http://www.youtube .com/watch?v=LVpGzCVQz4U.

Batty, David. "British Forces Complete Withdrawal from Basra." *The Guardian*, September 3, 2007. http://www.guardian.co.uk/world/2007/sep/03/military.iraq.

BBC. "Afghan Ambush Kills French Troops," August 19, 2008. http://news.bbc.co.uk /2/hi/7569942.stm.

———. "Forum Seeks to Ban Cluster Bombs," May 19, 2008. http://news.bbc.co.uk /2/hi/europe/7407631.stm.

———. "NATO's Afghan Forces 'Hit Limit,'" October 28, 2008. http://news.bbc.co .uk/2/hi/uk_news/7694350.stm.

———. "US Would Not 'Admit' the Insurgency in Post-War Iraq," December 15, 2009. http://news.bbc.co.uk/2/hi/uk_news/politics/8412317.stm.

———. "World Citizens Reject Torture, Global Poll Suggests," October 19, 2006. http://www.bbc.co.uk/pressoffice/pressreleases/stories/2006/10_october/19/poll .shtml.

Belasco, Amy. "Troop Levels in Afghanistan and Iraq Wars, FY 2001–FY2012: Cost and Other Potential Issues." Washington DC: Congressional Research Service, July 2, 2009. http://www.fas.org/sgp/crs/natsec/r40682.pdf.

Bellamy, Christopher. *The Evolution of Modern Land Warfare: Theory and Practice.* New York: Routledge, 1990.

Bender, Bryan, and Kevin Baron. "Fewer High-Quality Army Recruits: As War Needs Rise, Exam Scores Drop." *Boston Globe*, June 1, 2007. http://www.boston.com/news /nation/washington/articles/2007/06/01/fewer_high_quality_army_recruits/.

Benjamin, Daniel, and Steven Simon. *The Next Attack: Failure of the War on Terror and a Strategy for Getting It Right.* New York: Owl Books, 2006.

Bergen, Peter, Patrick Doherty, and Ken Bellen. "Public Opinion in Pakistan's Tribal Region." New America Foundation, September 28, 2010. http://www.newamerica .net/publications/policy/public_opinion_in_pakistan_s_tribal_regions.

Bially, Theodore. "Lifting the Fog of War." Presentation, *DARPATech*, Anaheim CA, August 9–11, 2005, 87–90. http://archive.darpa.mil/DARPATech2005 /presentations/ixo/bially.pdf.

Billings-Yun, Melanie. *Decision against War: Eisenhower and Dien Bien Phu, 1954.* New York: Columbia University Press, 1988.

Blank, Stephen. *Responding to Low-Intensity Conflict Challenges.* Maxwell AFB AL: Air University Press, 1990.

Blank, Stephen, Lawrence E. Grinter, Jerome W. Klingaman, Thomas P. Ofcansky, Lewis B. Ware, and Bynum E. Weathers. *Low-Intensity Conflict in the Third World.* Maxwell AFB AL: Air University Press, 1988.

Blankenship, Kaye. "Battlefield Ordnance Awareness (BOA)." Space and Missile Defense Command, U.S. Army. Presentation to the Army Science Board, June 2000. Document number RT T0600.082.

Blaufar, Douglas. *The Counterinsurgency Era: U.S. Doctrine and Performance, 1950 to Present.* New York: Free Press, 1977.

Boeing. "Apache Block III Reaches Another Milestone toward Production." Apache News, 2009. http://www.boeing.com/Microsites/ids/2009/apache/issue_01/news _s9_p2.html.

———. "Defense, Space and Security: AC-130U Gunship." Boeing Product Information, n.d. http://www.boeing.com/defense-space/support/maintenance/special _ops/ac130u.html.

———. "F-22 Raptor Specifications," n.d. http://www.lockheedmartin.com/us /products/f22/f-22-specifications.html.

———. "History: B-29 Superfortress," n.d. http://www.boeing.com/history/boeing /b29.html.

Bohan, Caren. "Obama Says Al Qaeda Still Greatest Threat to U.S." *Reuters*, November 16, 2009. http://www.reuters.com/article/idUSN16480901.

Borch, Fred. "Military Law and the Treatment of Women Soldiers: Sexual Harassment and Fraternization in the U.S. Army." In *A Soldier and a Woman: Sexual*

Integration in the Military, edited by Gerard J. De Groot and C. Peniston-Bird. Essex UK: Pearson Education Limited, 2000.

Bowden, Mark. *Black Hawk Down: A Story of Modern War*. New York: Atlantic Monthly Press, 1999.

Bowman, Tom, and Robert Siegel. "Examining U.S. Goals in Afghanistan." *All Things Considered*. National Public Radio, transcript, November 5, 2009. http://www.npr.org/templates/story/story.php?storyId=120138799.

Bradley, Mark Philip. *Vietnam at War*. Oxford: Oxford University Press, 2009.

Brady, Jeff. "Evangelical Chaplains Test Bounds of Faith in Military." *All Things Considered*. National Public Radio, transcript, July 27, 2005. http://www.npr.org/templates/story/story.php?storyId=4772331.

British 1st Airborne Division Living History Association. "Brief Battle History of the 2nd Parachute Battalion in World War II," 2005. http://www.1stairborne.com/history.html.

Brogan, Michael. "Marine Corps Force Protection Efforts." Statements for the House Armed Services Committee, February 4, 2009. http://armedservices.house.gov/pdfs/ALSPEF020409/Brogan_Testimony020409.pdf.

Brook, Tom Vanden. "Hulking, Tough MRAP Replaces Humvee." *USA Today*, March 23, 2010. http://www.usatoday.com/news/military/2010-03-22-humvee_N.htm.

———. "MRAPs Can't Stop Newest Weapon." *USA Today*, July 15, 2007. http://www.usatoday.com/news/world/iraq/2007-05-31-mrap-insurgents_N.htm.

Brower, Jennifer. "The Terrorist Threat and Its Implications for Sensor Technologies." In *Advances in Sensing with Security Applications*, edited by Jim Byrnes. Dordrecht, the Netherlands: Springer, 2005.

Browning Firearms. "A Brief History of the Browning BAR." *Browning eBlast Newsletter*, July 2008. http://www.browning.com/library/infonews/detail.asp?id=245.

Bull, Hedley. *The Anarchical Society: A Study of Order in World Politics*. 3rd ed. Basingstoke UK: Palgrave, 2002.

Burchett, Wilfred. *The Furtive War: The United States in Vietnam and Laos*. New York: International Publishers, 1963.

Burchill, Scott. *Theories of International Relations*. 2nd ed. New York: Palgrave, 2001.

Buttinger, Joseph. *Vietnam: A Dragon Embattled*. Vol. 2, *Vietnam at War*. New York: Praeger, 1967.

———. *Vietnam: A Political History*. New York: Praeger, 1968.

Buzan, Barry. *An Introduction to Strategic Studies: Military Technologies and International Relations*. New York: St. Martin's Press, 1987.

Cacciottolo, Mario. "Biometric Science Arrives at Heathrow." BBC, December 6, 2006. http://news.bbc.co.uk/2/hi/uk_news/6214592.stm.

Caldwell, Malcolm, and Lek Tan. *Cambodia in the Southeast Asian War*. New York: Monthly Review Press, 1973.

Callier, Maria. "Artificial Spider Silk Research Could Improve Body Armor, Parachutes." Wright-Patterson Air Force Base News, March 18, 2008. http://www.wpafb.af.mil/news/story.asp?id=123090646.

Callwell, C. E. *Small Wars: Their Principles and Practice*. London: Printed for His Majesty's Stationery Office by Harrison and Sons, 1896.

Carden, Michael J. "Iraqi Forces Improve as Violence Drops." American Forces Press Service, U.S. Central Command, August 8, 2008. http://www.centcom.mil/en /news/articles/iraqi-forces-improve-as-violence-drops.

Carew, Tom. *Jihad! The Secret War in Afghanistan*. Edinburgh UK: Mainstream, 2000.

Carlstorm, Gregg. "'Crazy Horse' and Collateral Damage: Helicopter Squadron that Killed Two Reuters Journalists in 2007 Was Involved in Other Attacks that Hurt Civilians." Al Jazeera, October 25, 2010. http://english.aljazeera.net/indepth /spotlight/2010/10/20101022165819333616.html.

Carl Zeiss Optronics. "Eyesafe Laser Rangefinder for the M1 Abrams Main Battle Tank." http://www.zeiss.com/c12571300034f59c/0/24b991a4d0d5d6dac125731500 4c8125/$file/53_0880e_m1.pdf.

Carollo, Russell. "Deadly Price Paid for Humvee Armor Used to Protect Soldiers." *Dayton Daily News*, June 11, 2006. http://www.daytondailynews.com/n/content /oh/story/news/special-reports/humvee/0611humvee.html.

Carroll, Joseph. "Slim Majority Supports Anti-Terrorism Action in Afghanistan, Pakistan: Public Continues to Support Military Efforts in Afghanistan." Gallup, August 8, 2007. http://www.gallup.com/poll/28333/slim-majority-supports -antiterrorism-action-afghanistan-pakistan.aspx.

Cecchine, Gary. *Army Medical Support to the Army after Next*. Santa Monica: RAND, 1999.

Cecchine, Gary, David Johnson, John Bondanella, J. Polich, and Jerry Sollinger. *Army Medical Strategy*. Santa Monica: RAND, 2001.

Central Intelligence Agency. "Country Comparison: Military Expenditures." *The World Factbook, 2005–6*. Washington DC: U.S. Government Printing Office, 2006. https://www.cia.gov/library/publications/the-world-factbook/rankorder /2034rank.html.

———. "Middle East: Iraq." *The World Factbook, 2010*. Washington DC: U.S. Government Printing Office, 2010. https://www.cia.gov/library/publications/the-world -factbook/geos/iz.html.

Chait, Richard, John Lyons, and Duncan Long. "Critical Technology Events in the Development of the Abrams Tank." Washington DC: Center for Technology and National Security Policy, National Defense University, December 2005. http:// www.dtic.mil/cgi-bin/GetTRDoc?ad=ada476340&Location=u2&doc=GetTR Doc.pdf.

Chandrasekaran, Rajiv. *Imperial Life in the Emerald City*. New York: Vintage Books, 2006.

———. "In Kandahar, a Taliban on the Rise." *Washington Post*, September 14, 2009. http://www.washingtonpost.com/wp-dyn/content/article/2009/09/13/ar20090 91302950.html.

Chesser, Susan. "Afghanistan Casualties: Military Forces and Civilians." Washington DC: Congressional Research Service, September 30, 2010. http://www.fas.org/sgp /crs/natsec/r41084.pdf.

Christian, Shirley. *Nicaragua: Revolution in the Family.* New York: Random House, 1985.

Clark, Wesley. *Winning Modern Wars: Iraq, Terrorism, and the American Empire.* New York: Public Affairs, 2003.

Clausewitz, Carl von. *On War.* Translated by Anatol Rapoport. Baltimore: Penguin Books, 1968.

CNN. "Pelosi Says No to Draft Legislation," November 20, 2006, http://articles.cnn .com/2006-11-20/politics/rangel.draft_1_selective-service-system-military -service-draft?_s=PM:POLITICS.

———. "Roadside Bombs 'No. 1 Threat' to Troops in Afghanistan," July 2009. http:// www.cnn.com/2009/WORLD/asiapcf/07/09/afghanistan.ieds/.

———. "White House: Murtha's Call Is 'Surrender,'" November 18, 2005. http:// articles.cnn.com/2005-11-17/politics/murtha.iraq_1_john-murtha-historic -democratic-election-draft-deferments?_s=pm:politics.

Cockburn, Patrick. "Sunni vs. Shia: The Real Bloody Battle for Baghdad." *The Independent*, February 5, 2008. http://www.independent.co.uk/news/world/middle -east/sunni-vs-shia-the-real-bloody-battle-for-baghdad-778038.html.

Cohen, Eliot, and Thomas Keaney. "Gulf War Air Power Survey: Summary Report." Washington DC: U.S. Government Printing Office, 1993. http://www.afhso.af .mil/shared/media/document/AFD-100927-061.pdf.

Cohen, Jon, and Jennifer Agiesta. "Poll of Afghans Shows Drop in Support for U.S. Mission." *Washington Post*, February 10, 2009. http://www.washingtonpost.com /wp-dyn/content/article/2009/02/09/AR2009020901368.html.

Coimbatore, Gopal, Steven M. Presley, Jonathan Boyd, Eric Marsland, and George P. Cobb. "Sensing Biological and Chemical Threat Agents." In *Advances in Biological and Chemical Terrorism Countermeasures*, edited by Ronald J. Kendall, Steven M. Presley, Galen P. Austin, and Philip N. Smith, 159–78. Boca Raton FL: CRC Press, 2008.

Combs, Cindy. "The Media as a Showcase for Terrorism." In *Teaching Terror: Strategic and Tactical Learning in the Terrorist World*, edited by James J. F. Forest, 133–54. New York: Rowman & Littlefield, 2006.

Comstock, Douglas, and Daniel Lockney. "NASA's Legacy of Technology Transfer and Prospects for Future Benefit." AIAA Space 2007 Conference and Exposition, Long Beach CA, September 18–20, 2007. http://www.sti.nasa.gov/tto/hist _techtransfer.pdf.

Cordesman, Anthony. *The Lessons of Afghanistan: War Fighting, Intelligence, and Force Transformation.* Washington DC: CSIS Press, 2002.

Cordovez, Diego, and Selig S. Harrison. *Out of Afghanistan: The Inside Story of the Soviet Withdrawal.* Oxford: Oxford University Press, 1995.

Council on Foreign Relations. "Q&A: What Is the Coalition of the Willing?" *New York Times International.* March 28, 2003. http://www.nytimes.com/cfr /international/slot1_032803.html.

Creveld, Martin van. *The Transformation of War.* New York: Free Press, 1991.

Crosby, John. "MRAPs Hits the Streets of Baquba." *The Desert Raider* 1, no. 7 (December 2007): 4, 5, 15.

Cubbage, Thomas, III. "Intelligence and the Tet Offensive: The South Vietnamese View of the Threat." In *The Vietnam War as History*, edited by Elizabeth Jane Errington and B. J. C. McKercher, 91. New York: Praeger, 1990.

Daily Mail, Foreign Service. "Video Footage Shows Taliban Fighters Swarming 'Valley of Death' Base Just Days after U.S. Troops Withdrawal." *Daily Mail*, April 20, 2010. http://www.dailymail.co.uk/news/worldnews/article-1267443 /Afghanistan-Taliban-fighters-swarm-US-military-base-days-troops-withdraw .html.

Dalloz, Jacques. *The War in Indo-China, 1945–54*. Dublin: Gill and MacMillan, 1990.

Damphousse, Kelly, and Brent Smith. "The Internet: A Terrorist Medium for the 21st Century." In *The Future of Terrorism: Violence in the New Millennium*, edited by Harvey W. Kushner, 208–24. London: SAGE Publications, 1998.

Darley, William M. "Clausewitz's Theory of War and Information Operations." *JFQ* 40 (2006): 73–79. http://www.au.af.mil/au/awc/awcgate/jfq/4015.pdf.

Davidson, Amy. "Sophisticated about Corruption?" *New Yorker*, September 16, 2010. http://www.newyorker.com/online/blogs/closeread/2010/09/sophisticated-about -corruption.html.

De Leon, Ruby. "Close NATO's 'Capability Gap' with Transatlantic Partnership." Speech, U.S. Department of Defense, June 26, 2000. http://www.defense.gov /speeches/speech.aspx?speechid=660.

DeLong, Michael. *Inside CENTCOM*. Washington DC: Regnery, 2004.

Denselow, Robin. "Agent Orange Blights Vietnam." BBC, December 3, 1998. http:// news.bbc.co.uk/1/hi/health/227467.stm.

Department of Defense. "Casualty Reports–Defenselink Casualty Report 2010." http://www.defense.gov/news/casualty.pdf.

———. "Operation Enduring Freedom Casualty Status." Department of Defense, Running Tally, n.d. http://www.defense.gov/news/casualty.pdf.

———. "Start of 'Arrowhead Ripper' Highlights Iraq Operation," June 19, 2007. http://www.defense.gov/news/newsarticle.aspx?id=46459.

Department of Defense, Joint Education and Doctrine Division, J-7, Joint Staff. *DOD Dictionary of Military Terms*. Online, 2009. http://www.dtic.mil/doctrine/jel /doddict/data/c/8822.html.

Dombrowski, Peter, and Eugene Gholz. *Buying Military Transformation: Technologi cal Innovation and the Defense Industry*. New York: Columbia University Press, 2006.

Dougherty, Jon. "Sabari: A Day in the Life of Missouri Combat Engineers." Missouri National Guard, June 23, 2010. http://www.ng.mil/news/archives/2010/06 /062410-Sabari.aspx.

Dougherty, P. J., and H. C. Eidt. "Wound Ballistics: Minié Ball vs. Full Metal Jacketed Bullets—a Comparison of Civil War and Spanish-American War Firearms." *Military Medicine* 174, no. 4 (2009): 403–7.

Drezner, Jeffery. *Innovative Management in the DARPA High Altitude Endurance Unmanned Aerial Vehicle Program.* Santa Monica: RAND, 1999.

Drummond, John. *Velocity Management: The Business Paradigm That Has Transformed U.S. Army Logistics.* Santa Monica: RAND, 2001.

Dubner, Stephen. "Would You Fly an Airplane with No Pilot?" *New York Times,* December 4, 2006.

Dupont. "Protecting Those Who Protect Us, with Body Armor,"2014. http://www.dupont.ca/en/products-and-services/personal-protective-equipment/body-armor.html.

Dupuy, Ernest, and Trevor Dupuy. *The Encyclopedia of Military History.* New York: Harper & Row, 1986.

Dupuy, Trevor. *The Evolution of Weapons and Warfare.* New York: Bobbs-Merill, 1980.

The Economist. Pocket World in Figures, 2002 Edition. Vicenza: L.E.G.O. S.P.A., 2002.

Ehrhart, Hans-Georg, and Charles Pentland. *The Afghanistan Challenge: Hard Realities and Strategic Choices.* Kingston, Canada: McGill-Queen's University Press, 2001.

Elliot-Bateman, Michael. *Defeat in the East: The Mark of Mao Tse-tung on War.* London: Oxford University Press, 1967.

Ellis, Chris, and Peter Chamberlain. *Fighting Vehicles.* New York: Hamlyn, 1972.

English, Richard. *Terrorism: How to Respond.* Oxford: Oxford University Press, 2009.

Fadel, Leila. "As U.S. Scales Back Role in Iraq, Attacks and Political Deadlock Persist." *Washington Post,* August 22, 2010. http://www.washingtonpost.com/wp-dyn/content/article/2010/08/21/ar2010082102383.html.

Fall, Bernard B. *Hell in a Very Small Place: The Siege of Dien Bien Phu.* London: Pall Mall, 1967.

―――. *Street without Joy.* 4th ed. Harrisburg PA: Stackpole, 1964.

―――. *Viet-Nam Witness, 1953–66.* New York: Praeger, 1966.

Federation of American Scientists (FAS), Military Analysis Network. "Laser Guided Bombs," February 2000. http://www.fas.org/man/dod-101/sys/smart/lgb.htm.

―――. "Non-lethal and Riot Control Agents." Chemical Weapons Convention Archive, n.d. http://www.fas.org/blog/cw/document-archive/documents-by-subject/non-lethal-and-riot-control-ag.

Feickert, Andrew. "The Army's Future Combat System (FCS): Background and Issues for Congress." Washington DC: Congressional Research Service, May 12, 2008. http://fpc.state.gov/documents/organization/106166.pdf.

Finel, Bernard I. "An Alternative to COIN." *Armed Forces Journal,* 2007. http://www.afji.com/2010/02/4387134.

Florer, Hayward. "A New Phase in the War against Terrorism Begins as U.S. Special Operations Forces Land in Afghanistan." Public Broadcasting Service, October 19, 2001.

Fotion, Nicholas. *War and Ethics: A New Just War Theory.* Think Now. New York: Continuum, 2007.

Fox News. "Transcript: Gen. David Petraeus on 'Fox News Sunday,'" December 23, 2007. http://www.foxnews.com/story/0,2933,318049,00.html.

Fox, Stephen. "JSTARS Adds Blue Force Tracking Capability." U.S. Air Force News, January 19, 2006. http://www.af.mil/news/story.asp?storyID=123014563.

Franco, Horacio. SRI Systems, quoted December 19, 2005. http://www.uscg.mil/hr /cgi/downloads/Military_dot_Com_Article.pdf.

Frankland, Noble, ed. *The Encyclopedia of Twentieth-Century Warfare*. New York: Mitchell Beazley Publishers, 1989.

Franks, Tommy. *American Soldier*. With Malcolm McConnell. New York: Regan Books, 2004.

Fredricksson, Brian. "Space Power in Joint Operations: Evolving Concepts." *Aerospace Power Journal* 18, no. 2 (Summer 2004).

Freeman, Douglas Southall. *Lee's Lieutenants: A Study in Command*. New York: Charles Scribner's Sons, 1942.

Friedman, George, and Meredith Friedman. *The Future of War: Power, Technology & American World Dominance in the Twenty-first Century*. New York: St. Martin's Griffin, 1998.

Fuller, J. F. C. *The Conduct of War, 1789–1961: A Study of the Impact of the French, Industrial, and Russian Revolutions on War and Its Conduct*. New Brunswick NJ: Rutgers University Press, 1961.

Fussell, Paul. *The Great War and Modern Memory*. New York: Oxford University Press, 1975.

Gaddis, John Lewis. *Surprise, Security, and the American Experience*. Cambridge MA: Harvard University Press, 2004.

Galeotti, Mark. *The Soviet Union's Last War*. London: Frank Cass, 1995.

Galula, David. *Counterinsurgency Warfare: Theory and Practice*. PSI Classics of the Counterinsurgency Era. Westport CT: Praeger Security International, 2006.

Gebhardt, James. "Eyes behind the Lines: U.S. Army Long-Range Reconnaissance and Surveillance Units." Global War on Terrorism Occasional Paper 10, rev. ed. Fort Leavenworth KS: Combat Studies Institute Press, October 31, 2005. http:// www.cgsc.edu/carl/download/csipubs/gebhardt_LRRP.pdf.

Gelpi, Christopher, Peter D. Feaver, and Jason Reifler. *Paying the Human Costs of War: American Public Opinion and Casualties in Military Conflicts*. Princeton NJ: Princeton University Press, 2009.

General Dynamics Canada. "Fire Control," n.d.

Gerring, George. "The Vietnam Analogy and the 'Lessons' of History." In *The Vietnam War as History*, edited by Elizabeth Jane Errington and B. J. C. McKercher. New York: Praeger, 1990.

Gibbons, Meghan. "Representing the Real on the Road to Guantanamo." In *The War on Terror and American Popular Culture: September 11 and Beyond*, edited by Matthew Hill and Andrew Schopp. Cranbury NJ: Associated University Press, 2010.

Gilbert, Marc Jason. *The Vietnam War: Teaching Approaches and Resources*. New York: Greenwood Press, 1991.

Gillespie, Joan. *Algeria: Rebellion and Revolution*. New York: Praeger, 1960.

Girardet, Edward. *Afghanistan: The Soviet War*. New York: St. Martin's Press, 1985.

Global Security. "Appendix C: Target Analysis and Munition Effects and Terminal Ballistics," May 13, 2010. http://www.globalsecurity.org/military/library/policy /army/fm/6-40/Appc.htm.

———. "Military: MC-130P Combat Shadow," 2010. http://www.globalsecurity.org /military/systems/aircraft/mc-130p.htm.

———. "OIF-Iraq Significant Activities (SIGACTS)," May 13, 2010. http://www .globalsecurity.org/military/ops/iraq_sigacts.htm.

Goghlan, Tom. "American Troops Pull Out of Korengal Valley as Strategy Shifts." *The Times*, April 15, 2010.

Goldberg, Bernard. *Arrogance: Rescuing America from the Media Elite.* New York: Warner Books, 2003.

———. *Bias: A CBS Insider Exposes How the Media Distort the News.* Washington DC: Regnery, 2001.

Goldman, Marshall. *Gorbachev's Challenge: Economic Reform in the Age of High Technology.* New York: W. W. Norton, 1987.

Goodson, Larry. *Afghanistan's Endless War: State Failure, Regional Politics, and the Rise of the Taliban.* Seattle: University of Washington Press, 2001.

Gordon, Michael R., and Andrew W. Lehren. "Leaked Reports Detail Iran's Aid for Iraq Militias." *New York Times*, October 22, 2010. http://www.nytimes.com/2010 /10/23/world/middleeast/23iran.html.

Gordon, Michael, and Bernard E. Trainor. *The Generals' War: The Inside Story of the Conflict in the Gulf.* New York: Little, Brown, 1995.

Gray, Colin S. *The Sheriff: America's Defense of the New World Order.* Lexington: University Press of Kentucky, 2004.

———. *Strategy for Chaos: Revolutions in Military Affairs and the Evidence of History.* Strategy and History Series. London: Frank Cass, 2002.

———. *Transformation and Strategic Surprise.* Carlisle Barracks PA: Strategic Studies Institute, U.S. Army War College, 2005.

Gray, J. Glenn. *The Warriors: Reflections on Men in Battle.* Lincoln: University of Nebraska Press, 1998.

Greenfield, Kent Roberts, ed. *Command Decisions.* Washington DC: Center of Military History United States, Department of the Army, 1960.

Gudmundsson, Bruce I. *Stormtroop Tactics: Innovation in the German Army, 1914–1918.* London: Praeger, 1989.

Haas, Richard. *War of Necessity, War of Choice: A Memoir of Two Iraq Wars.* New York: Simon & Schuster, 2009.

Haddick, Robert. "Learning from the Korengal Valley." *Small Wars Journal*, April 14, 2010. http://smallwarsjournal.com/blog/2010/04/learning-from-the-korengal -val/.

———. "This Week at War: The Long Death of the Powell Doctrine." *Foreign Policy*, March 5, 2010. http://www.foreignpolicy.com/articles/2010/03/05/this_week_at _war_the_powell_doctrine_is_dead.

Handel, Michael I. *Sun Tzu and Clausewitz: The Art of War and on War Compared.* Professional Readings in Military Strategy no. 2. Carlisle Barracks PA: Strategic Studies Institute, U.S. Army War College, 1991.

Hansen, Harry. *The Civil War: A History.* New York: Penguin Books, 1991.

Hansen, John, Umit Yapanel, Rongqing Huang, and Ayako Ikeno. "Dialect Analysis and Modeling for Automatic Classification." Department of Electrical Engineering, University of Texas, 2004. Presented at INTERSPEECH 2004, Eighth International Conference on Spoken Language Processing, Jeju Island, Korea, October 4–8, 2004. See http://www.researchgate.net/publication/221488981_Dialect_analysis_and_modeling_for_automatic_classification.

Harringon, Nancy, and Edward Doucette. "Army after Next and Precision Airdrop." *U.S. Army,* 1999. http://www.almc.army.mil/alog/issues/JanFeb99/MS388.htm.

Harrington, Nathan. "Patent Application Title: Method and System for Vehicle Traffic Monitoring Based on the Detection of a Characteristic Radio Frequency." Federation of American Scientists Patent Monitoring, November 27, 2008. http://www.faqs.org/patents/app/20080291055.

Harrison, Jeff. "Americans Turn to Al-Jazeera for Raw Images of War, UA Study Finds." *UA News,* April 18, 2010. http://uanews.org/node/31241.

Hashim, Ahmed. *Iraq's Sunni Insurgency.* Adelphi Paper 402. London: International Institute for Strategic Studies, 2009.

Hastings, Max. *The Oxford Book of Military Anecdotes.* New York: Oxford University Press, 1985.

Henry, Kim. "UAS VideoTerminal Connects Boots on the Ground to Eyes in the Sky." Space War, UAV News, October 1, 2007. http://www.army.mil/article/5148/uas-video-terminal-connects-boots-on-the-ground-to-eyes-in-the-sky/.

Herper, Mathew. "LifePoint Makes New Portable Drug Test." *Forbes,* October 9, 2000. http://www.forbes.com/2000/10/10/1010ftech.html.

Herrington, Stuart A. *Silence Was a Weapon: The Vietnam War in the Villages.* Novato CA: Presidio Press, 1982.

Hirsch, John L., and Robert B. Oakley. *Somalia and Operation Restore Hope: Reflections on Peacemaking and Peacekeeping.* Washington DC: United States Institute of Peace, 1995.

Hoar, Joseph. "Why Aren't There Enough Troops in Iraq?" *New York Times.* Op-ed, April 2, 2003.

Hung, Nguyen. "The Vietnam War in Retrospect: Its Nature and Some Lessons." In *The American War in Vietnam: Lessons, Legacies, and Implications for Future Conflicts,* edited by Lawrence E. Grinter and Peter M. Dunn. New York: Greenwood Press, 1987.

Hurley, William. *Taking the "Revolution in Military Affairs" Downtown: New Approaches to Urban Operations.* Alexandria VA: Institute for Defense Analysis, 2001.

International Committee of the Red Cross. "Convention (IV) relative to the Protection of Civilian Persons in Time of War, 12 August 1949," Article 33. http://www

.icrc.org/ihl.nsf/385ec082b509e76c41256739003e636d/6756482d86146898c12564
1e004aa3c5.

Jackson, Randy, and Chris Haddox. "Phantom Eye High Altitude Long Endurance
Aircraft Unveiled." Boeing, June 12, 2010. http://www.boeing.com/Features/2010
/07/bds_feat_phantom_eye_07_12_10.html.

Jaffe, Greg. "U.S. Retreat from Afghan Valley Marks Recognition of Blunder." *Wash-ington Post*, April 15 2010. http://www.washingtonpost.com/wp-dyn/content/
article/2010/04/14/AR2010041401012.html.

Jane's Electro-Optic Systems. "BAE Systems AN/PAS-13C-V Thermal Weapon Sight
II," September 10, 2007. http://www.janes.com/articles/Janes-Electro-Optic-Systems/BAE-Systems-AN-PAS-13C-V-Thermal-Weapon-Sight-II-TWS-II
-United-States.html.

Johns, David. "The Crimes of Saddam Hussein: Suppression of the 1991 Uprising."
Frontline World. Public Broadcasting Service, 1999. http://www.pbs.org/frontline
world/stories/iraq501/events_uprising.html.

Johnson, David, and Gary Cecchine. *Medical Risk in the Future Force Unit of Action:
Results of the Army Medical Department Transformation Workshop IV*. Santa
Monica: RAND, 2005.

Jones, Sam. "Obama's Afghanistan Deadline Gives Taliban Sustenance, US General
Warns." *The Guardian*, August 25, 2010. http://www.guardian.co.uk/world/2010
/aug/25/obama-afghanistan-deadline-sustains-taliban.

Jordan, David. *Understanding Modern Warfare*. New York: Cambridge University
Press, 2008.

Junger, Sebastian. "Farewell to Korengal." *New York Times*, April 20, 2010. http://
www.nytimes.com/2010/04/21/opinion/21junger.html.

———. *War*. New York: Grand Central Publishing, 2010.

Kahib, George. *Intervention: How America Became Involved in Vietnam*. New York:
Alfred A. Knopf, 1986.

Keaney, Thomas. "Lessons Unlearned: The History of Airpower in Unconventional
Warfare." *Armed Forces Journal* (online), June 1, 2006. http://www.armedforces
journal.com/lessons-unlearned/.

Keiler, Jonathan. "Who Won the Battle of Fallujah?" *Proceedings*, January 2005.
http://www.military.com/NewContent/1,13190,NI_0105_Fallujah-P1,00.html.

Keller, Hans W. "The Piezoresistive Technology on the Right Track," 1999. http://
www.keller-druck.com/home_e/painfo_e/berichte_2000_e.asp but the date
= March 2000.

Kepel, Gilles, and Jean-Pierre Milelli. *Al Qaeda in Its Own Words*. Cambridge MA:
The Belknap Press of Harvard University Press, 2008.

Khong, Yuen Foong. *Analogies at War: Korea, Munich, Dien Bien Phu, and the Vietnam
Decisions of 1965*. Princeton NJ: Princeton University Press, 1992.

Kikuchi, Shigeo. "Iraq Troop Surge of 2007 and the U.S. Civil-Military Relations."
The National Institute for Defense Studies News, no. 133 (June 2009). http://www
.nids.go.jp/english/publication/briefing/pdf/2010/133.pdf.

Kilcullen, David. *The Accidental Guerrilla: Fighting Small Wars in the Midst of a Big One*. New York: Oxford University Press, 2009.

Kinross, Stuart. *Clausewitz and America: Strategic Thought and Practice from Vietnam to Iraq*. New York: Routledge, 2008.

Kirby, Alex. "Depleted Uranium Ban Demanded." BBC, December 17, 1999. http://news.bbc.co.uk/2/hi/science/nature/568234.stm.

Kissinger, Henry, and Clare Boothe Luce. *White House Years*. Boston: Little, Brown, 1979.

Kitson, Frank. *Bunch of Five*. London: Faber, 1977.

Klein, Joe. "The Pashtun War," *Time*, May 5, 2009. http://swampland.blogs.time.com /2009/05/05/the-pashtun-war/.

Kline, Harvey. *Colombia: Democracy under Assault*. Boulder CO: Westview Press, 1995.

Knight, Ben. "Amnesty International Accuses Israel of War Crimes in Gaza." ABC News, January 20, 2009. http://www.abc.net.au/news/stories/2009/01/20 /2469609.htm.

Komarow, Steven. "Unexpected Insurgency Changed Way of War," *USA Today*, March 20, 2005, http://www.usatoday.com/news/world/iraq/2005-03-20-change -war_x.htm.

Kopp, Carlo. "JDAM Matures Parts 1 and 2." *Australian Aviation*, August 2008. http://www.ausairpower.net/TE-JDAMPt1.html.

Kramer, Don. "4th Brigade's Deployment Adjusted for Early Autumn." Army.mil, March 5, 2009. http://www.army.mil/-news/2009/03/05/17841-4th-brigades -deployment-adjusted-for-early-autumn/.

Krepinevich, Andrew. *The Army and Vietnam*. Baltimore MD: Johns Hopkins University Press, 1986.

Krishnamachari, Bhaskar. *Networking Wireless Sensors*. Cambridge: Cambridge University Press, 2005.

Krishnan, Armin. *Killer Robots: Legality and Ethicality of Autonomous Weapons*. Surrey, UK: Ashgate Publishing, 2009.

Kristol, William. "The Defense Secretary We Have." *Washington Post*, December 15, 2004. http://www.washingtonpost.com/wp-dyn/articles/A132-2004Dec14.html.

Kull, Steven. "Negative Attitudes toward the United States in the Muslim World: Do They Matter?" World Public Opinion.org, May 17, 2007. Transcript of congressional testimony.

Lambeth, Benjamin. *Air Power against Terror: America's Conduct of Operation Enduring Freedom*. Santa Monica: RAND, 2005.

Lang, W. Patrick, Jr. "What Iraq Tells Us about Ourselves." *Foreign Policy*, February 15, 2007. http://www.foreignpolicy.com/articles/2007/02/14/what_iraq_tells_us _about_ourselves.

Law, John, and Annemarie Mol. *Complexities: Social Studies of Knowledge Practices*. Durham NC: Duke University Press, 2002.

Lawson, Alastair. "Pakistan's Islamic Schools in the Spotlight." BBC, July 14 2005. http://news.bbc.co.uk/2/hi/south_asia/4683073.stm.

Leckie, Robert. *The Wars of America*. New York: Harper & Row, 1967.

Legal Information Institute, Cornell University Law School. U.S. Code, Title 10—Armed Forces, Subtitle A—General Military Law, Part II—Personnel, Chapter 47—UCMJ, Subchapter 10—Punitive Articles. http://www.law.cornell.edu/uscode/html/uscode10/usc_sup_01_10_10_A_20_ii_30_47_40_X.html.

———. U.S. Code, Title 18—Crimes and Criminal Procedures, Part I—Crimes, Chapter 50A—Genocide. http://www.law.cornell.edu/uscode/18/1091.html.

Lehner, Paul. *Artificial Intelligence and National Defense: Opportunity and Challenge*. Blue Ridge Summit PA: TAB Books.

Lesser, Ian O., Bruce Hoffman, John Arquilla, David Ronfeldt, Michele Zanini, and Brian Michael Jenkins. *Countering the New Terrorism*. Santa Monica: RAND, 1999.

Levinger, Josh. "Taking Tora Bora," May 6, 2005. http://www.levinger.net/josh/files/writing/tora-bora.pdf.

Library of Congress, Business Reference Services. "Modern Automobile Manufacturing." *BERA: Business & Economics Research Advisor* 2 (Fall 2004). http://www.loc.gov/rr/business/BERA/issue2/manufacturing.html.

Lockheed Martin. "Lockheed Martin Receives First Contract for LOSAT Production Missiles." Press release, August 9, 2002. http://www.lockheedmartin.com/news/press_releases/2002/LockheedMartinReceivesFirstContract.html.

———. "Lockheed Martin's LOSAT Defeats Moving Target in White Sands Missile Range Test." Press release, August 8, 2003. http://www.lockheedmartin.com/news/press_releases/2003/LockheedMartinSLOSATDefeatsMovingTa.html.

———. "Lockheed Martin's LOSAT Successfully Fires Guided Missile during Engineering Development Flight Test." Press release, June 11, 2003. http://www.lockheedmartin.com/news/press_releases/2003/LockheedMartinSLOSATSuccessfullyFir.html.

Lock-Pullan, Richard. "U.S.: Military Strategy, Strategic Culture, and the War on Terror." In Owens and Dumbrell, *America's "War on Terrorism."*

Loeb, Vernon. "Rumsfeld Untracks 'Crusader'; Weapon's End Faces Hill GOP Challenge." *Washington Post*, May 9, 2002. http://www.washingtonpost.com/ac2/wp-dyn?pagename=article&node=&contentId=A53762-2002may8&notFound=true.

Lombardo, Tony. "Army to Field Upgraded Raven UAV in December." *Army Times*, November 2, 2009. http://www.armytimes.com/news/2009/11/army_raven_110209w/.

Luttwak, Edward. *Coup d'État: A Practical Handbook*. Cambridge MA: Harvard University Press, 1979.

Lynn, John A. "Patterns of Insurgency and Counterinsurgency." *Military Review*, July–August 2005. http://www.au.af.mil/au/awc/awcgate/milreview/lynn.pdf.

MacAskill, Evan, and Julian Borger. "Iraq War Was Illegal and Breached UN Charter, Says Annan." *The Guardian*, September 16, 2004. http://www.guardian.co.uk/world/2004/sep/16/iraq.iraq.

Macgregor, Douglas A. *Breaking the Phalanx: A New Design for Landpower in the 21st Century*. Westport CT: Praeger, 1997.

Mahalik, Nitaigour P. *Sensor Networks and Configuration: Fundamentals, Standards, Platforms, and Applications*. Berlin: Springer, 2007.

Mao Zedong and Samuel B. Griffith. *Yu Chi Chan (Guerrilla Warfare)*. Norfolk VA: Reproduced by G-2 Headquarters, Fleet Marine Force, Atlantic, 1961.

Margolis, Eric. *War at the Top of the World: The Struggle for Afghanistan, Kashmir, and Tibet*. New York: Routledge, 2000.

Marshall, S. L. A. "Big Guns Not Answer." In *Vietnam: Anatomy of a Conflict*, edited by Welsley Fishel. Itasca IL: F. E. Peacock Publishers, 1968.

———. *The Soldier's Load and the Mobility of a Nation*. Quantico VA: Marine Corps Association, 1980.

Martone, James. "Questions over Afghan Arabs." CNN, December 20, 2001. http://archives.cnn.com/2001/WORLD/meast/12/19/arabs.in.afghanistan/.

Matsumura, John, Randall Steeb, Thomas Herbert, Scot Eisenhard, John Gordon IV, Mark Lees, and Gail Halverson. "The Army after Next: Exploring New Concepts and Technologies for the Light Battle Force." Santa Monica: RAND, 1997. http://www.rand.org/pubs/documented_briefings/DB258/DB258.pdf.

McAlster, John, Jr. *Vietnam: The Origins of Revolution*. New York: Alfred A. Knopf, 1969.

McCurry, Justin. "Rice Says Sorry for U.S. Troop Behavior on Okinawa as Crimes Shake Alliance with Japan." *The Guardian*, February 28, 2008. http://www.guardian.co.uk/world/2008/feb/28/japan.usa.

McFadden, Cynthia, and Ammu Kannampilly. "Clinton Nears Decision of U.S. Strategy in Afghanistan." *Nightline*. ABC News, Oct 14, 2009. http://abcnews.go.com/Nightline/International/hillary-clinton-nears-decision-us-strategy-afghanistan/story?id=8815655.

McMichael, William H. "Hold Fire, Earn a Medal." *Navy Times*, May 11, 2010. http://www.navytimes.com/news/2010/05/military_restraint_medal_051110mar/.

Megargee, Geoffery. *War of Annihilation: Combat and Genocide on the Eastern Front*. Oxford: Rowman & Littlefield, 2006.

Melton, Jack W., and Lawrence E. Pawl. *Melton & Pawl's Guide to Civil War Artillery Projectiles*. Kennesaw GA: Kennesaw Mountain Press, 1996.

Michaels, Jim. "Petraeus Strategy Takes Aim at Post-Vietnam Mind-set." *USA Today*, March 8, 2007. http://www.usatoday.com/news/world/iraq/2007-03-07-petraeus_N.htm.

Millen, Raymond. "Tweaking NATO: The Case for Integrated Multinational Divisions." Carlisle Barracks PA: Strategic Studies Institute, U.S. Army War College, June 2002. http://www.strategicstudiesinstitute.army.mil/pdffiles/pub130.pdf.

Miller, Jon. "Hugo Grotius." In *Stanford Encyclopedia of Philosophy*, vol.1, edited by Edward N. Zalta. Stanford CA: Metaphysics Research Lab, Center for the Study of Language and Information, Stanford University, December 16, 2005. http://plato.stanford.edu/entries/grotius/.

Ministry of Foreign Affairs, Embassy of the Islamic Republic of Afghanistan–Warsaw. "Afghanistan: Ethnic Groups," 2013. http://www.afghanembassy.com.pl/eng/afganistan/grupy-etniczne.

Molnar, Andrew. "Human Factors Considerations of Undergrounds in Insurgencies." 2nd ed. With Jerry M. Tinker and John D. LeNoir. Washington DC: Special Operations Research Office, American University, December 1, 1965. http://www.cgsc.edu/carl/docrepository/dapam550_104insurgencies.pdf.

Monbiot, George. "Behind the Phosphorus Clouds Are War Crimes within War Crimes." *The Guardian*, November 22, 2005. http://www.guardian.co.uk/world/2005/nov/22/usa.iraq1.

Monks, Alfred. *The Soviet Intervention in Afghanistan*. Washington DC: American Enterprise Institute for Public Policy Research, 1981.

Montross, Lynn. *War through the Ages*. Rev. and enl. 3rd ed. New York: Harper, 1960.

Moore, Harold G., and Joseph L. Galloway. *We Were Soldiers Once . . . and Young: Ia Drang—the Battle That Changed the War in Vietnam*. New York: Random House, 1992.

Morrison, Chris. "A Sonic Silver Bullet for Fighting Crime." CNN Money, November 9, 2009. http://money.cnn.com/2009/11/09/smallbusiness/ear_for_crime.fsb/index.htm.

Mosely, Phillip. *The Kremlin and World Politics*. New York: Vintage Books, 1960.

Mrozek, Donald. *Air Power and the Ground War in Vietnam: Ideas and Actions*. Maxwell AFB AL: Air University Press, 1988.

Nagl, John A. *Learning to Eat Soup with a Knife: Counterinsurgency Lessons from Malaya and Vietnam*. Chicago: University of Chicago Press, 2002, 2005.

Nardulli, Bruce R., and Thomas L. McNaugher, "The Army: Toward the Objective Force." In *Transforming America's Military*, edited by Hans Binnendijk, 101–28. Washington DC: National Defense University Press, 2002.

NASA (National Aeronautics and Space Administration). "Innovative Partnership Program," June 9, 2009. http://www.nasa.gov/offices/ipp/home/myth_tools.html.

National Oceanic and Atmospheric Association, National Climatic Data Center. "Climate of Afghanistan," August 20, 2008. http://www.ncdc.noaa.gov/oa/climate//afghan/afghan-narrative.html.

National Research Council. *Decadal Survey of Civil Aeronautics: Foundations for the Future*. Washington DC: National Academies Press, 2006.

Naylor, Sean. *Not a Good Day to Die: The Untold Story of Operation Anaconda*. New York: Berkley Books, 2005.

Newell, Nancy Peabody, and Richard S. Newell. *The Struggle for Afghanistan*. Ithaca NY: Cornell University Press, 1981.

Nikutta, Randolph. "Artificial Intelligence and the Automated Tactical Battlefield." In *Arms and Artificial Intelligence: Weapons and Arms Control Applications of Advanced Computing*, edited by Allan M. Din. New York: Oxford University Press, 1987.

North Atlantic Treaty Organization (NATO). "Frequently Asked Questions: What Does NATO Do?," March 11, 2009. http://www.nato.int/cps/en/natolive/faq.htm.

Northrup Grumman Aerospace Systems. "MQ-8B Fire Scout," 2012. http://www.northropgrumman.com/Capabilities/FireScout/Documents/pageDocuments/MQ-8B_Fire_Scout_Data_Sheet.pdf.

Northrup Grumman Electronic Systems. "AN/AAQ-24(V) Directional Infrared Countermeasures (DIRCM): In Production, Deployed and Tailored to Any Platform," n.d. http://www.es.northropgrumman.com/solutions/nemesis/.

Nossal, Kim Richard. "No Exit: Canada and the 'War without End' in Afghanistan." In *The Afghanistan Challenge: Hard Realities and Strategic Choices*, edited by Hans-Georg Ehrhart and Charles C. Pentland, 157–73. Kingston, Canada: McGill-Queens' University Press, 2009.

O'Connell, John. *The Effectiveness of Airpower in the 20th Century*. New York: iUniverse, 2006.

Operation Iraqi Freedom, Official Website of United States Forces Iraq. "Diyala," July 31, 2009.

Owens, Bill. *Lifting the Fog of War*. With Ed Offley. New York: Farrar, Straus and Giroux, 2000.

Owens, John, and John Dumbrell. *America's "War on Terrorism": New Dimensions in U.S. Government and National Security*. Plymouth UK: Lexington Books, 2008.

Paddock, Alfred, Jr. *U.S. Army Special Warfare: Its Origins*. Lawrence: University of Kansas Press, 2002.

Padover, Saul Kussiel. *The Living U.S. Constitution*. 2nd ed. Revised by Jacob W. Landynski. New York: New American Library, 1983.

Paget, Julian. *Counterinsurgency Operations: Techniques of Guerrilla Warfare*. New York: Walker and Company, 1967.

Paschall, Rod. *The Defeat of Imperial Germany, 1917–1918*. New York: Da Capo Press, 1994.

Patai, Raphael. *The Arab Mind*. New York: Charles Scribner's Sons, 1973.

Patton, George S. *War as I Knew It*. Annotated by Paul D. Harkins. Boston: Houghton Mifflin, 1995.

Penfold, Chuck. "German Chancellor Defends Afghanistan Mission." *Deutsche Welle*, March 22, 2010. http://www.dw-world.de/dw/article/0,,5491899,00.html.

Perez-Davis, Marla, Patricia L. Loyselle, Mark A. Hoberecht, Michelle A. Manzo, Lisa L. Kohout, Kenneth A. Burke, and Carlos R. Cabrera. "Energy Storage for Aerospace Applications." Proceedings of 36th Intersociety Energy Conversion and Engineering Conference, July–August 2001. http://gltrs.grc.nasa.gov/reports/2001/TM-2001-211068.pdf.

Peterson, Harold. *A History of Body Armor*. New York: Charles Scribner's Sons, 1968.

Petraeus, David, and James Amos. *FM 3-24: Counterinsurgency*. Washington DC: Headquarters, Department of the Army, 2006.

Pierson, Ruth. "They're Still Women After All: Wartime Jitters over Femininity." In *Women and the Military System*, edited by Eva Isaksson. New York: St. Martin's Press, 1988.

Poole, H. J. *One More Bridge to Cross: Lowering the Cost of War*. Emerald Isle NC: Posterity Press, 1999.

Prados, John. *The Hidden History of the Vietnam War*. Chicago: Ivan Dee, 1995.

———. *Vietnam: The History of an Unwinnable War, 1945–1975*. Lawrence: University Press of Kansas, 2009.

Prokosch, Eric. *The Technology of Killing: A Military and Political History of Antiper-sonnel Weapons*. London: Atlantic Highlands NJ: Zed Books, 1995.

Public Broadcasting Service. "Interview: Lt. Col. Ernest 'Rock' Marcone." *Frontline*, 2004. http://www.pbs.org/wgbh/pages/frontline/shows/invasion/interviews /marcone.html.

Putsay, John. *Counterinsurgency Warfare*. New York: Free Press, 1965.

Quester, George. "The Problem." In *Female Soldiers—Combatants or Noncombatants? Historical and Contemporary Perspectives*, edited by Nancy L. Goldman. London: Greenwood Press, 1982.

Rais, Rasul Bakhsh. *War without Winners: Afghanistan's Uncertain Transition after the Cold War*. Karachi: Oxford University Press, 1994.

RAND. "Taliban's Sanctuary Bases in Pakistan Must Be Eliminated." News release, June 9, 2008. http://www.rand.org/news/press/2008/06/09/.

Rashid, Ahmed. *Descent into Chaos: The United States and the Failure of Nation Building in Pakistan, Afghanistan, and Central Asia*. London: Viking Penguin, 2008.

Raytheon. "Raytheon's APG-79 AESA Radar for the F/A-18 Super Hornet Sets a New Standard as It Delivers Multiple JDAMs Simultaneously on Target." Raytheon Media Relations, December 5, 2005. http://raytheon.mediaroom.com /index.php?s=43&item=329&pagetemplate=release.

Reeve, Simon. *The New Jackals: Ramzi Yousef, Osama bin Laden, and the Future of Terrorism*. Boston: Northeastern University Press, 1999.

Rhem, Kathleen T. "Bush: No Distinction between Attackers and Those Who Harbor Them." U.S. Department of Defense, American Forces Press Service, September 11, 2001. http://www.defense.gov/news/newsarticle.aspx?id=44910.

Ricks, Thomas E. *The Gamble: General David Petraeus and the American Military Adventure in Iraq, 2006–2008*. New York: Penguin Press, 2009.

———. "Rumsfeld Gets Earful from Troops." *Washington Post*, December 9, 2004. http://www.washingtonpost.com/wp-dyn/articles/A46508-2004Dec8.html.

Rincon, Paul. "Super-Strong Body Armor in Sight." BBC, October 23, 2007. http:// news.bbc.co.uk/2/hi/science/nature/7038686.stm.

Roggio, Bill. "Iranian-Backed Shia Terror Group Remains a Threat in Iraq: General Odierno." *The Long War Journal*, July 13, 2010. http://www.longwarjournal.org /archives/2010/07/iranianbacked_shia_t_1.php.

Rosenau, William. "Special Operations Forces and Elusive Enemy Ground Targets." RAND Project AIR FORCE, 2002. http://www.rand.org/pubs/research_briefs/RB 77/index1.html.

Rosenberg, Matthew. "U.S. Forces Leave Afghan 'Valley of Death.'" *Wall Street Journal*, April 15, 2010. http://online.wsj.com/article/SB10001424052702304159304575 183383654837248.html.

Rothstein, Hy. *Afghanistan and the Troubled Future of Unconventional Warfare*. Annapolis MD: Naval Institute Press, 2006.

Rubin, Barnett. *The Fragmentation of Afghanistan*. New Haven CT: Yale University Press, 1995.

————. *The Search for Peace in Afghanistan: From Buffer State to Failed State*. New Haven CT: Yale University Press, 1995.

Ruelle, David. *Chance and Chaos*. Princeton NJ: Princeton University Press, 1991.

Saad, Lydia. "Americans' Opinion of UN at Record Low." Gallup, March 6, 2008. http://www.gallup.com/poll/104806/americans-opinion-un-record-low.aspx.

Sarin, Oleg, and Lev Dvoretsky. *The Afghan Syndrome: The Soviet Union's Vietnam*. Novato CA: Presidio, 1993.

Satterfield, John. "A Long and Winding Military Procurement Road: The V-22 Osprey Story." *National Examiner*, May 20, 2009. http://www.examiner.com /national-security-in-national/a-long-and-winding-military-procurement-road -the-v-22-osprey-story.

Scales, Robert H. *Yellow Smoke: The Future of Land Warfare for America's Military*. Lanham MD: Rowman & Littlefield, 2003.

Schabas, William. "Convention for the Prevention and Punishment of the Crime of Genocide: Paris, 9 December 1948." Codification Division, Office of Legal Affairs, United Nations, 2008. http://legal.un.org/avl/pdf/ha/cppcg /cppcg_e.pdf.

Scheuer, Michael. *Through Our Enemies' Eyes*. Washington DC: Brassey's, 2002.

Schmidt, Eric. "Pentagon Contradicts General on Iraq Occupation Force Size." *New York Times*, February 28, 2003. http://www.globalpolicy.org/component/content /article/167/35435.html.

Schmidtchen, David. *Eyes Wide Open: Stability, Change and Network-Enabling Technology*. Land Warfare Studies Centre Working Paper Series. Duntroon, Australia: Land Warfare Studies Centre, May 2006.

Schmitt, Gary. "Defense Reorganization and the Office of Net Assessment." Project for the New American Century, November 10, 1997.

Schwarzkopf, H. Norman. *It Doesn't Take a Hero: General H. Norman Schwarzkopf, the Autobiography*. With Peter Petre. New York: Bantam Books, 1992.

Shank, Patti. *The Online Learning Idea Book: 95 Proven Ways to Enhance Technology Based and Blended Learning*. San Francisco: John Wiley and Sons, 2007.

Sharp, U. S. Grant. Jr. *Strategy for Defeat: Vietnam in Retrospect*. San Rafael CA: Presidio Press, 1978.

Shay, Jonathan. *Achilles in Vietnam: Combat Trauma and the Undoing of Character*. New York: Simon & Schuster, 1994.

Shinseki, Eric, U.S. Army Chief of Staff. "Concepts for the Objective Force." White Paper. Washington DC: U.S. Army, 2001.

Shinwari, Ibrahim. "Pakistanis Reopen Khyber Pass for Afghan Supplies." Reuters, January 2, 2009. http://www.reuters.com/article/idUSTRE5011N620090102.

Simonite, Tom. "'Robot Arms Race' Underway, Experts Warn." *New Scientist*, February 27, 2008. http://www.newscientist.com/article/dn13382-robot-arms-race -underway-expert-warns.html.

Simpson, Howard. *Dien Bien Phu: The Epic Battle America Forgot*. Washington DC: Brassey's, Inc., 1994.

Slack, James. "Ministers Accused of Trying to Build DNA Database by Stealth." *Daily Mail*, September 5, 2007. http://www.dailymail.co.uk/news/article-480017 /Ministers-accused-trying-build-DNA-database-stealth.html.

Smith, Roger M. *Cambodia's Foreign Policy*. Ithaca NY: Cornell University Press, 1965.

Sorley, Lewis. *A Better War: The Unexamined Victories and Final Tragedy of America's Last Years in Vietnam*. New York: Harcourt Brace, 1999.

Starr, Harvey. *War Coalitions: The Distributions of Payoffs and Losses*. London: Lexington Books, 1972.

Starry, Donn A. *Armored Combat in Vietnam*. New York: Arno Press, 1980.

Staugaard, Andrew. *Robotics and AI: An Introduction to Applied Machine Intelligence*. Englewood Cliffs NJ: Prentice-Hall, 1987.

Steele, Brett. "Military Reengineering between the World Wars." Santa Monica: RAND, 2005, 59–60. http://www.rand.org/pubs/monographs/2005/RAND _MG253.pdf.

Steen, Donald Von. "Remote Sensing for Crop and Livestock Estimates." Washington DC: Statistical Reporting Service, U.S. Department of Agriculture, 1968. http://www.nass.usda.gov/research/reports/Internet.

Steinhauser, Paul. "CNN Poll: Afghanistan War Opposition at All-Time High." CNN, September 1, 2009. http://politicalticker.blogs.cnn.com/2009/09/01/cnn-poll -afghanistan-war-opposition-at-all-time-high/.

Stratemeyer, George E., and William T. Y'Blood. *The Three Wars of Lt. Gen. George E. Stratemeyer: His Korean War Diary*. Washington DC: Air Force History and Museums Program, 1999.

Straub, Amanda. "Female Combat Medics Earn Respect for Afghan Army." *U.S. Army News*, January 29, 2007. http://www.army.mil/-news/2007/01/29/1550 -female-combat-medics-earn-respect-from-afghan-army/.

Sturgill, Claude. *Low-Intensity Conflict in American History*. London: Praeger, 1993.

Surlis, Paul. "Iraq War, Unjust, Illegal and Immoral; Just War Theory Condemns Invasion." *Houston Catholic Worker* 25, no. 1 (January–February 2005). http://cjd. org/2005/02/01/iraq-war-unjust-illegal-and-immoral-just-war-theory -condemns-invasion/.

Swartz, Jon, and Edward Iwata. "Invented to Save Gas, Kevlar Now Saves Lives." *USA Today*, April 15, 2003. http://www.usatoday.com/money/world/iraq/2003-04 -15-kevlar_x.htm.

Taber, Robert. *War of the Flea: The Classic Study of Guerrilla Warfare*. Washington DC: Brassey's, Inc., 2002.

Tanner, Stephen. *Afghanistan: A Military History from Alexander the Great to the Fall of the Taliban*. New York: Da Capo Press, 2002.

Taylor, William, S. Craig Moore, and Charles Robert Roll, Jr. *The Air Force Pilot Shortage: A Crisis for Operational Units?* Santa Monica: RAND, 2000.

ThinkExst.com Quotations. "Napoleon Bonaparte Quotes," 2014. http://thinkexist .com/quotation/a_soldier_will_fight_long_and_hard_for_a_bit_of/216923.html.

Thistlewaite, Susan Brooks, Rev., and Brian Katulis. "How to Make Afghanistan War a 'Just War.'" Center for American Progress, November 19, 2009. http://www.americanprogress.org/issues/2009/11/just_war.html.

Thompson, Robert Grainger Ker. *Defeating Communist Insurgency: Experiences from Malaya and Vietnam*. Studies in International Security no. 10. London: Chatto & Windus, 1966.

Thompson, W. Scott, and Donaldson D. Frizzell. *The Lessons of Vietnam*. New York: Crane, Russak, 1977.

Thro, Ellen. *Robotics: The Marriage of Computers and Machines*. New York: Facts on File, 1993.

Thucydides. *The Peloponnesian War*. Chicago: William Benton, 1952.

Tolz, Vera. "Independence for Russian Minorities: War in Chechnya." In *The Collapse of the Soviet Union*, edited by Paul A. Winters. San Diego CA: Greenhaven Press, 1999.

Travers, Tim. *The Killing Ground: The British Army, the Western Front, and the Emergence of Modern War, 1900–1918*. South Yorkshire: Pen and Sword Military Classics, 1987.

Tucker, David. "Confronting the Unconventional: Innovation and Transformation in Military Affairs." Letort Papers, no. 14. Carlisle Barracks PA: Strategic Studies Institute, U.S. Army War College, 2006.

Tucker, David, and Christopher Lamb. *United States Special Operations Forces*. New York: Columbia University Press, 2007.

Tyson, Ann Scott. "Sharp Drop in Gays Discharged from the Military Tied to War Need." *Washington Post*, March 14, 2007. http://www.washingtonpost.com/wp-dyn/content/article/2007/03/13/AR2007031301174.html.

Tzu, Sun. *The Art of War*. Translated by Lionel Giles. Philadelphia PA: Miniature Editions, 2003.

Ucko, David. *The New Counterinsurgency Era: Transforming The U.S. Military for Modern Wars*. Washington DC: Georgetown University Press, 2009.

United Nations. "Convention on Prohibitions or Restrictions on the Use of Certain Conventional Weapons Which May Be Deemed to Be Excessively Injurious or to Have Indiscriminate Effects," October 13, 1995. http://www.un.org/millennium/law/xxvi-18-19.htm.

————. "Convention on the Prohibition of the Use, Stockpiling, Production and Transfer of Anti-Personnel Mines and on Their Destruction," September 18, 1997. http://www.un.org/Depts/mine/UNDocs/ban_trty.htm.

United States Army Special Operations Command. "A Brief History of Special Operations Forces." http://www.soc.mil/sofinfo/history.html.

United States General Accounting Office (GAO). "Combat Air Power: Funding Priority for Suppression of Enemy Air Defenses May Be Too Low." GAO/NSIAD-96-128. Washington DC: GAO, April 10, 1996. http://www.globalsecurity.org/military/library/report/gao/ns96128.pdf.

———. "Military Transformation: Army Has a Comprehensive Plan for Managing Its Transformation but Faces Major Challenges." GAO-02-96. Washington DC: GAO, November 2001, 1. http://www.gao.gov/new.items/d0296.pdf.

Urban, Mark. *War in Afghanistan*. New York: St. Martin's Press, 1990.

U.S. Army. "Chapter 5: Soldier's Load Management and Training for FOOT MARCHES." *Field Manual 21-18: Foot Marches*. Washington DC: Department of the Army, June 1990. https://rdl.train.army.mil/soldierPortal/atia/adlsc/view/public/9499-1/fm/21-18/FM211_6.htm.

———. "5th Special Forces Group (Airborne): History." Fort Campbell, n.d. http://www.campbell.army.mil/units/5thSFG/Pages/5thGroup.aspx.

———. "Frequently Asked Questions about Recruiting," October 13, 2010.

———. *Introduction to Special Forces Operations*. U.S. Army John F Kennedy Special Warfare Center and School. Fort Bragg NC: U.S. Army, 1993.

———. "Introduction to Special Forces Operations" [course]. Ft. Bragg NC: JFK Special Warfare Center and School, 1993.

———. "A Leader's Guide to After-Action Reviews." Training Circular 25-20. Washington DC: Headquarters, Department of the Army, September 30, 1993. http://www.au.af.mil/au/awc/awcgate/army/tc_25-20/tc25-20.pdf.

———. "Operational Army: The Objective Force." Posture statement, 2003. http://www.army.mil/aps/2003/realizing/transformation/operational/objective/index.html.

———. "Special Forces: The Missions of Special Forces," n.d.

———. "Special Forces: Training Overview," n.d.

U.S. Marine Corps. *Command and Control*. Washington DC: Department of the Navy, 1996.

Valeriano, Napolean, and Charles Bohannan. *Counter-Guerrilla Operations: The Philippine Experience*. New York: Praeger, 1962.

Van Tien Dung. *Our Great Spring Victory: An Account of the Liberation of South Vietnam*. New York: Monthly Review Press, 1977.

Vick, Alan, Adam Grissom, William Rosenau, Beth Grill, and Karl P. Mueller. *Air Power in the New Counterinsurgency Era: The Strategic Importance of USAF Advisory and Assistance Missions*. Santa Monica: RAND, 2006.

Vickers, Michael G., and Robert C. Martinage. "The Revolution in War." Washington DC: Center for Strategic and Budgetary Assessments, 2004.

Viola, James A. "What Is the Proper Role of Public Opinion in the Decision to Use Military Force as an Element of National Power." Carlisle Barracks PA: U.S. Army War College, March 15, 2006. http://ics.leeds.ac.uk/papers/pmt/exhibits/2967/ksil525.pdf.

Wailes, William, and Nancy Harrington. "The Guided Parafoil Airborne Delivery System Program." Reston VA: American Institute of Aeronautics and Astronautics, 2004.

Walker, David. "DOD Transformation: Challenges and Opportunities." Washington DC: U.S. Government Accountability Office," February 12, 2007. http://www.gao.gov/cghome/d07500cg.pdf.

Wall Street Journal. "Al Qaeda in Africa: Sunday's Uganda Attacks Underscore a New Global Threat," July 13, 2010. http://online.wsj.com/article/SB1000142405274870 4288204575363071312256954.html.

Walt, Stephen. "There Are No 'Do-Overs' in Long Wars." *Foreign Policy*, September 7, 2010. http://walt.foreignpolicy.com/posts/2010/09/07/no_do-overs_in _long_wars.

Waxman, Matthew. "International Law and the Politics of Urban Air Operations." Santa Monica: RAND, 2000. http://www.rand.org/pubs/monograph_reports/ MR1175.html.

Webster, George. "Biometric ATM Gives Cash via 'Finger Vein' Scan." CNN, July 5, 2010. http://articles.cnn.com/2010-07-05/world/first.biometric.atm.europe_1 _banking-markets-polish-bank-bank-cards?_s=PM:WORLD.

Welch, Michael. *Crimes of Power and States of Impunity: The U.S. Response to Terror.* London: Rutgers University Press, 2009.

Wells, Matt. "Al-Jazeera Accuses U.S. of Bombing Its Kabul Office." *The Guardian*, November 17, 2001. http://www.guardian.co.uk/media/2001/nov/17/warin afghanistan2001.afghanistan.

West, Bing. *The Strongest Tribe: War, Politics, and the Endgame in Iraq.* New York: Random House, 2009.

West, Mike. "Minie Balls Were Battlefield Revolution." *Murfreesboro Post*, September 2, 2009. http://www.murfreesboropost.com/news.php?viewStory=1778.

Wilmoth, Gregory. "False-Failed Innovation." *Joint Forces Quarterly*, Autumn/Winter 1999/2000. http://www.dtic.mil/doctrine/jfq/jfq-23.pdf.

Windrow, Martin. *The Last Valley.* Cambridge MA: Da Capo Press, 2006.

Winters, Harold A. *Battling the Elements: Weather and Terrain in the Conduct of War.* With Gerald E. Galloway Jr., William J. Reynolds, and David W. Rhyne. Baltimore: Johns Hopkins University Press, 1998.

Woodward, Bob. *Obama's Wars.* New York: Simon & Schuster, 2010.

———. *Plan of Attack.* New York: Simon & Schuster, 2004.

Woodward, Peter. *U.S. Foreign Policy and the Horn of Africa.* Chippenham: Ashgate, 2006.

World Public Opinion. "International Poll: No Consensus on Who Was Behind 9/11." WorldPublicOpinion.org, September 10, 2008. http://www.worldpublic opinion.org/pipa/articles/international_security_bt/535.php.

Wright, Evan. *Generation Kill: Devil Dogs, Iceman, Captain America and the New Face of American War.* New York: Putnam's Sons, 2004.

Yoo, John. *War by Other Means: An Insider's Account of the War on Terror.* New York: Atlantic Monthly Press, 2006.

Yousaf, Mohammad, and Mark Adkin. *Afghanistan: The Bear Trap: The Defeat of a Superpower.* Barnsley UK: Leo Cooper, 1992.

Zabecki, David. *Steel Wind: Colonel Georg Bruchmüller and the Birth of Modern Artillery.* London: Praeger, 1994.

Zumbro, Ralph. *The Iron Cavalry.* New York: Simon & Schuster, 1998.

Index

Mundweil, Phillip, 89–90
Murtha, John, 208

Nagl, John, 6, 12, 14–15, 92
Napoleon, 188
NASA-developed technology, spin-offs from, 137–38
national will. *See* public opinion (U.S.)
NATO, 49, 126, 196–98
Navarre, Henry, 41, 42
NCOs, professionalism of, 219–20
neighborhood watch programs in Iraq, 88–89
net-centric warfare, 144–45

Obama administration: Afghanistan and, 110–12, 126–31, 177; Iraq and, 207
Office of Net Assessment, of U.S. Defense Department, 92–93
operationally offensive, tactically defensive concept, 25–66; coalition partners, 194–99; foundation and risks of, 25–28; future and, 60–66; infantry and, 56–60; Iraq and, 96–99, 100; limits to U.S. options in Afghanistan, 108, 109, 115, 123; maintaining national will, 206–11, 217–18, 231; need for sound strategy, 157–58; proposed deployment in villages, 158–66; use of term, 3
operationally offensive, tactically defensive concept and military technology: air power, 43–56; artillery, 32–38; UAVs and UCAVs, 166–69
operationally offensive, tactically defensive concept historically: Dien Bien Phu, 27, 38–45, 51–53, 134; Gettysburg, 28–33, 35, 44, 133; Khe Sanh, 50–56, 134
operationally offensive, tactically defensive concept and insurgency: counterinsurgency manpower, 183–89; insurgents' countering of U.S. tactics, 199–206; insurgents' tactics

and strategy and, 173–79; insurgents' tactics and strategy, countering of, 179–83; insurgents' third-party safe havens, 191–94; insurgents' third-party technical support, 189–91
Operation Market Garden, in World War II, 27, 31

Pakistan: Soviet Union in Afghanistan and, 119–22, 193; United States in Afghanistan and, 125–26, 127
Pashtun people. *See* Afghanistan and U.S. efforts to end Taliban support for al-Qaeda
Patton, George, 29
Petraeus, David, 6, 12–13, 88, 96, 97, 103, 146, 207
Piroth, Charles, 42
populace in area of insurgency: and change in military mindset and dispersal of infantry, 58–60; winning hearts and minds of, 6–7, 11, 12, 15. *See also* public opinion (U.S.)
Powell, Colin, 192
precision-guided munitions, 149; in Afghanistan, 123–24; in Iraq, 74–78
propaganda, insurgent tactics and, 201–6
public opinion (U.S.): counterinsurgency and maintaining of national will, 206–11, 217–18, 231; enemy tactics and strategy and, 175–76. *See also* populace in area of insurgency

radio frequency (RF) sensors, 152, 154
Reagan administration, 119
reconnaissance and operationally offensive, tactically defensive concept, 182
resupply of counterinsurgency forces, 154–55, 162–66
"revolution in military affairs": concerns about technology transferred to adversaries, 139–40; evolutionary, not revolutionary change and, 135–37;

implications of, 226–27; products building on multiple technologies, 137–39, 140–41; use of term, 2–3

risk aversion as counterproductive, 217

Rorke's Drift, battle of (1879), 27

Rumsfeld, Donald, 72, 141, 172

rural areas, sensor technology's use in, 151–52

safe havens, third-party providing for insurgents, 191–94

San Francisco, sensor technology used in, 151, 153

Scales, Robert, 16, 20

sensor technology, 16, 174; and communications in proposed counterinsurgency strategy, 158–62; in operationally offensive, tactically defensive proposal, 157–69, 227–28, 230; usefulness and drawbacks of, 150–54

September 11 attacks (2001), varied public reactions to, 203–4, 210

Shadow UAV, 83–86, 98

Siler, David, 84

soda straw effect of sensor information, 154

Somalia, 71, 106

Soviet Union and counterinsurgency choices in Afghanistan, 116–22, 193

space. See time, trading for space

space-based sensing and observation, role in counterinsurgency, 149

spades as Civil War technology, 28, 30, 31, 32–33, 35

state as political unit of importance to United States, 7, 192

Steele, Brett, 19, 136–37, 141

Stinger surface-to-air missiles and U.S. and Soviet differences in Afghanistan, 120–22, 124–125, 171, 190

Stratemeyer, George, 46–47

Stryker combat vehicle, 81

Sun Tzu, 6, 7–8, 9, 11, 28, 65

surface-to-air missiles and U.S. and Soviet differences in Afghanistan, 120–22, 124–125

Taber, Robert, 6, 12, 14, 213, 214

Taliban. See Afghanistan and U.S. efforts to end Taliban support for al-Qaeda

technical asymmetry: decline in infantry numbers and, 58–60; operationally offensive, tactically defensive concept and, 27–28, 31, 65, 218

technology, generally: implications for, 225–28; questions of appropriate use of, 3–5; strategy and, 17–19; third-party support for insurgents and, 189–91. See also counterinsurgency strategy, technology and

Thompson, Robert, 6, 12, 14, 213

time, trading for space, and operationally offensive, tactically defensive concept, 35–36, 38, 53–54, 64–65, 150

Tora Bora, battle of, 52

Tucker, David, 15

unconventional warfare. See insurgency warfare

United Kingdom, 198, 231

United Parcel Service (UPS), Worldport facility of, 165–66

United States: changing the narrative option in Afghanistan, 100, 111–12, 126–28, 131; realist tradition and national goals of, 6–7; technical advantages over Soviet Union in Afghanistan, 122–26; unrestricted operations option in Afghanistan, 111–12, 129–31, 224–25; withdrawal option in Afghanistan, 110–11, 126, 128–29, 177, 224

unmanned aerial vehicles (UAVs): in Iraq, 82–86; potential uses for, 156–57; in proposed counterinsurgency strategy, 166–69 unmanned combat air vehicles (UCAVs), 152, 154; potential uses for, 156–57; in proposed counterinsurgency strategy, 166–69

unrestricted operations option for counterinsurgency without local support: generally, 106–7, 113–16; United States in Afghanistan and, 111–12, 129–31, 224–25

urban areas, sensor technology's use in, 151

U.S. Army, "revolution in military affairs" and, 135–37

U.S. Army Special Forces, 98, 186; need for complementary conventional forces, 142–44

U.S. military: lessons from Iraq and, 219–20; proper counterinsurgency practice and, 216–19. *See also specific units*

vehicles (land) in Iraq, 80–82

Vickers, Michael, 3

Vietnam: Mao's three stages of revolutionary warfare, 173–74; misuse of technology in, 17–18, 215. *See also* Dien Bien Phu; Khe Sanh

Vinh Yên, battle of, 40

withdrawal option for counterinsurgency without local support: generally, 106, 107–9; United States in Afghanistan and, 110–11, 126, 128–29, 177, 224

women, needed in counterinsurgency work, 186–87

Worldport facility of United Parcel Service (UPS), 165–66

World War I: *grignotage* (nibbling) in, 31; trench warfare and operationally offensive, tactically defensive concept, 33–38, 44, 56

World War II, German blitzkrieg in, 18

Yellow Smoke (Scales), 16

Yousaf, Mohammad, 119